Target Recognition and Tracking for Millimeter Wave Radar in Intelligent Transportation

Lin Cao · Zongmin Zhao · Dongfeng Wang

Target Recognition and Tracking for Millimeter Wave Radar in Intelligent Transportation

中国工信出版集团 電子工業出版社 PUBLISHING HOUSE OF ELECTRONICS INDUSTRY http://www.phei.com.cn

Springer

Lin Cao
School of Information and Communication Engineering
Beijing Information Science and Technology University
Beijing, China

Zongmin Zhao
School of Information and Communication Engineering
Beijing Information Science and Technology University
Beijing, China

Dongfeng Wang
Research and Development Department
Beijing TransMicrowave Technology Co., Ltd.
Beijing, China

This work was supported by National Science Foundation of China (U20A20163, 62001034), Scientific Research Project of Beijing Municipal Education Commission (KZ202111232049, KM202011232021) and Qin Xin Talents Cultivation Program (QXTCP A201902).

ISBN 978-981-99-1532-3 ISBN 978-981-99-1533-0 (eBook)
https://doi.org/10.1007/978-981-99-1533-0

Jointly published with Publishing House of Electronics Industry
The print edition is not for sale in China (Mainland). Customers from China (Mainland) please order the print book from: Publishing House of Electronics Industry.
ISBN of the Co-Publisher's edition: 978-712-14-2091-7

This Springer imprint is published by the registered company Springer Nature Singapore Pte Ltd.
The registered company address is: 152 Beach Road, #21-01/04 Gateway East, Singapore 189721, Singapore

Preface

In the past 20 years, the field of intelligent transportation has developed rapidly, and it has continued to progress from nothing to new. Nowadays, intelligent transportation equipment can be seen everywhere. The application of intelligent transportation equipment makes people's travel planning more convenient. Judging from the current traffic situation, the development of intelligent transportation field is far from meeting people's normal travel needs. It is necessary to establish more complete laws and regulations, form more complete industry norms, design more efficient intelligent traffic control systems, and develop more sophisticated edge node devices.

At present, the main traffic monitoring method adopts the combination of radar and camera that the radar detects the target vehicle, following by capturing relevant information with the camera. In the recent years, image processing technologies represented by machine learning and deep learning have attracted a lot of attention, while the development of civilian radars has rarely been mentioned. Compared with cameras, millimeter-wave radar has huge advantages. It has a wide monitoring range and a long distance. It is easy to achieve a sense of over-the-horizon, and the amount of uploaded data is small, which reduces the system load. More importantly, it has a strong anti-interference ability, which is not afraid of the influence of climate and environment. Therefore, we believe that the deep integration of image processing technology and millimeter-wave radar data processing technology is important content to promote the development of the field of intelligent transportation.

By reading this book, readers can understand the role of millimeter-wave radar in the field of intelligent transportation, as well as the architecture of the vehicle tracking system based on MIMO millimeter-wave radar, and the processing method of millimeter-wave radar detection data. From the point of view of the radar, to understand the whole system in depth, it is necessary to know not only what it is, but also why it is. The book consists of 7 chapters. Chapter 1 introduces the application and research situation of millimeter-wave radar in the field of intelligent transportation, as well as radar data and image processing techniques; Chap. 2 introduces several commonly used traffic radar systems and compares them; Chap. 3 introduces new denoising algorithms based on traditional denoising algorithms and introduces a method of correcting the radar installation swing angle using measurement data; Chap. 4 introduces the detection and tracking technology of vehicles in practical scenes based on the measured data; Chap. 5 introduces the classical clustering algorithm and proposes the spindle-based density peak fuzzy clustering; Chap. 6 introduces the technology of target tracking using data association algorithm and the design of multi-target tracking system; and Chap. 7 briefly introduces image processing technology and the fusion method of image and radar data.

Based on my own experience, the relevant knowledge of millimeter-wave radar data processing in intelligent transportation is combed and summarized in this book. Hope to play a role in attracting jade and promoting the development of relevant industries. Through summarization, I can also deeply understand the overall structure and improve my theoretical level. The publication of this book has been supported by the National Natural Science Foundation of China (Project No.: U20A20163, 62001033, 61671069), the Beijing Municipal Education Commission Science and Technology Program Project (Project No.: KZ202111232049, KM202011232021, KM202111232014), and Beijing Information Science and Technology University "Qinxin Talents" Cultivation Program. I would like to express my thanks for the funding of scientific research projects such as the project (Project No.: QXTCP A201902, C202108).

Due to the limited level of the author, there are inevitably deficiencies in the book, and readers please understand and correct.

Beijing, China
October 2021

Lin Cao

Contents

Chapter 1
Introduction

1.1 Research Background and Significance

In the 1990s, intelligent transportation industry started, and gradually transformed from tracking international advanced technologies to developing core technologies with independent intellectual property rights. Its rapid development has brought a huge impact on the society. Traffic information is the basic information of urban traffic planning and management. Obtaining comprehensive, rich and real-time traffic information can not only grasp the development status of urban roads, but also predict the development situation and provide a scientific basis for the correct decision-making of traffic management departments.

At present, the collection of traffic information is mainly completed by radar, camera, sonar, etc. The distance, speed, angle and other information of the measured target can be obtained by processing the information.

In the field of transportation, there are many solutions based on video image information. Factors such as lighting will cause great interference to the video image information, so these solutions cannot be adapted to all-weather intelligent traffic scenarios. In the highway traffic scene, the accident rate at night is high, the traffic supervision has problems such as unclear vision and difficulty in determining responsibility, and the effect of manual supervision and video supervision is weak. Millimeter-wave radar can be well adapted to complex scenes, and has the advantages of all-weather and strong environmental adaptability. More and more researchers begin to use radar to collect traffic information.

Millimeter-wave radar is a detection radar with wavelengths greater than microwaves and less than centimeter waves, and its operating frequency is 30 to 300 GHz. The main features of millimeter-wave radar are strong reliability, high detection accuracy, strong stability, all-weather, strong anti-interference ability, convenient installation and maintenance, long life, and low initial installation cost. Millimeter-wave radar is widely used in the military field because of its high resolution, strong anti-interference ability (environment, illumination, etc.), small size, and all-weather work. In recent years, civilian millimeter-wave radars have continued to

L. Cao et al., *Target Recognition and Tracking for Millimeter Wave Radar in Intelligent Transportation*, https://doi.org/10.1007/978-981-99-1533-0_1

develop, such as traffic surveillance radars, vehicle-mounted radars, and fixed-point ranging radars.

Most of the early research on millimeter-wave radar target tracking systems was aimed at a one-to-one system. In a one-to-one system, the radar can only track the trajectory of one target. With the development of related technologies and the complexity of the scene, many scholars began to study the multi-target tracking system. In 1955, Wax proposed the basic concept of multi-target tracking. In multi-target tracking, clustering algorithm, data association algorithm and filtering algorithm are extremely important. In 1984, Aldenderfer et al. proposed four functions of cluster analysis: first, further expansion of data classification; second, conceptual exploration of entity classification; third, generation of hypotheses through data exploration; fourth, a method based on actual data sets. How the categorization hypothesis is tested. Sittler put forward the concept of the target point trace and the optimal data association of the track. The filtering algorithm was not used in the tracking field at first, and the filtering algorithm was used for target tracking after 1970. The most classic filtering algorithm is the linear filtering proposed by Kalman in 1960.

In today's rapid development of science and technology, safety is an important issue that we cannot ignore. Millimeter-wave radar can be used to track multiple targets and predict vehicle trajectories for passing vehicles, and to provide early warning of dangerous driving behaviors, thereby effectively reducing the occurrence of traffic accidents and ensuring traffic safety.

Using radar for multi-target tracking is a comprehensive application, and the entire tracking process requires a lot of theoretical and algorithmic support. The flow of multi-target tracking using radar is shown in Fig. 1.1.

Using radar for multi-target tracking involves radar signal processing theory, radar data processing theory, clustering algorithm theory, data association algorithm theory, and filtering algorithm theory. In the highway traffic scene, there are some adjacent vehicles whose echo signals are similar and occlude each other, which affects the target tracking results. Based on the relevant data collected by the millimeter-wave radar in the traffic scene, this book introduces the millimeter-wave radar target tracking algorithm, and introduces the clustering algorithm and data association algorithm in detail.

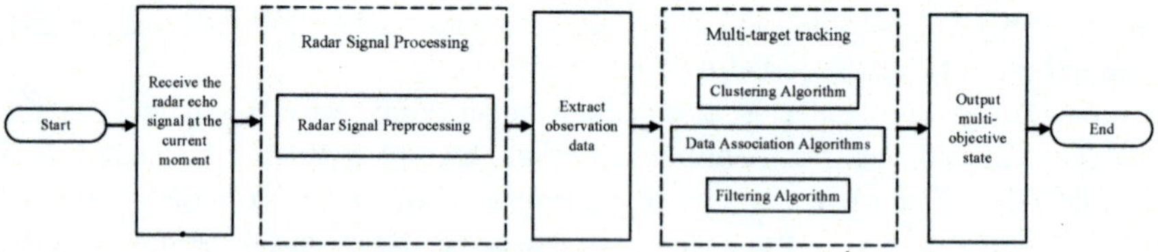

Fig. 1.1 Flow of multi-target tracking using radar

1.2 Research Status

1.2.1 Research Status of Traffic Surveillance Radar

At present, traffic surveillance radar is mainly used in high-speed road condition monitoring and urban traffic light intelligent control. It can detect illegal parking, speeding, illegal merging, emergency lane driving and other events within 250 m ahead, obtain the position of each vehicle in the current surveillance scene and track all the trajectory of the vehicle. Users can get real-time traffic information through radar, so as to identify traffic jams, abnormalities and other events. Traffic surveillance radar should include all or part of the following functions: ① The radar realizes multi-lane detection, automatically divides lanes, tracks vehicles in different lanes, and accurately detects the speed and position of vehicles; ② It can detect the oncoming vehicle and the going vehicle separately, and it can also detect the oncoming and going vehicle at the same time; ③ According to the vehicle trajectory, it can judge abnormal events such as reverse driving, emergency stop, abnormal lane change, rear-end collision, collision, etc.; horizontal coverage of 1 to 10 Lane, covering 200–300 m longitudinally, realizes large scene detection; ④ Contains the function of traditional microwave vehicle detector; ⑤ It can judge the congestion information of the lane, count the length of the congested vehicles, and send the congestion information; ⑥ The resolution of distance and speed is high, the vehicle positioning is accurate, and the speed measurement is accurate; ⑦ Self-contained real-time clock, which can be synchronized with the network clock; ⑧ Self-contained storage, data will not be lost when power off.

There are many mature radar products abroad. For example, Germany's Smartmicro's traffic information collection radar, the radar's name is Universal Medium Range Radar (UMRR). The company has also designed a suite of application software that provides good feedback on the information collected by the radar. When the user opens the software to run the adapted radar, he can get the return information from the software interface, including the target, lane, monitoring range, etc. The entire system is composed of radar and supporting software, which can provide information such as the position, speed and vehicle type of each monitored target, and use the target detection line to count traffic flow information and display the vehicle's driving trajectory. UMRR is shown in Fig. 1.2. The US-based Wavetronix has also designed the Smart Sensor HD radar with superior performance. The radar has a detection distance of 76 m and a detection width of 10 lanes. The official traffic flow detection accuracy rate is over 95%, and the speed measurement accuracy rate is about 97%.

At the beginning of the twenty-first century, China began to develop traffic surveillance radars. At present, the more mature products are mainly speed measuring radars, and comprehensive traffic surveillance radars are still in the research and development stage. Beijing Chuansu Microwave Technology Co., Ltd. is a leader in the research and development of millimeter-wave radar for intelligent transportation in

Fig. 1.2 UMRR

China. The microwave speed measuring radar developed by the company has excellent performance, can accurately monitor the speeding behavior of vehicles, and trigger the camera to capture, with a success rate of more than 99%.

CSR-TS radar is a new generation of traffic surveillance radar independently developed by Beijing Chuansu Microwave Technology Co., Ltd. It adopts international advanced microwave radar technology and has made important breakthroughs in radar system, signal processing, target recognition algorithm, etc. CSR-TS radar can be widely used in high-speed road condition monitoring and intelligent control of traffic lights. The application scenarios include road condition detection of highways and urban highways, warning of wrong-way traffic and illegal lane changes, and providing information such as queue length for intelligent control of traffic lights at intersections. The main functions of the radar are: it can monitor 4 to 5 lanes at the same time, and the number of monitoring lanes is adjustable; it has high speed and ranging resolution; it can know the direction of the vehicle speed; the track tracking accuracy rate exceeds 90%, and the tracking distance accuracy is ± 1 m; Real-time monitoring and judgment of wrong-way vehicles, lane-changing vehicles, and emergency-stop vehicles; judging the status of traffic jams and queues in the scene. The installation position of the CSR-TS radar is usually the center position in the width direction of the multi-lane. In order to expand the application scope of radar monitoring and reduce the installation requirements, it can be properly deviated from this position during actual installation. In addition to top mounting, there are also radar mounting methods such as side top mounting and side mounting. The CSR-TS radar is shown in Fig. 1.3.

The main technical parameters and main performance indicators of CSR-TS radar are shown in Tables 1.1 and 1.2 respectively.

1.2.2 *Research Status of Radar Data Processing*

There are two important processes for multi-target tracking with radar in traffic scenarios: clustering and data association. The purpose of clustering is to extract valid

Fig. 1.3 CSR-TS radar

Table 1.1 Main technical parameters of CSR-TS radar

Technical parameter	Attributes
Antenna type	Flat microstrip array antenna
Antenna beamwidth	Emission: 8° × 12°@3 dB (Far) 8° × 20°@3 dB (Middle) 8° × 80°@3 dB (Close) Accept: 8° × 80°@3 dB
Center frequency	76.5 GHz
Center frequency error	$\leq \pm 10$ MHz
Transmit power	$\leq$ 13 dBm
Power	9 ~ 36 V
Rated power	6 W
Range of working temperature	− 40 ~ 70°C
Operating humidity range	5 ~ 95% RH
Radar size	147 mm × 84 mm × 48 mm
Communication interface	RS485, network port
Installation method	Top mount, side top mount, side mount

observations and complete the classification, and the purpose of data association is to correctly associate the observations with the predicted points of the target. The observation point is mainly composed of sample point data obtained from radar echo signals, including target information and noise signals. The prediction point is the target position estimate at the next moment obtained by predicting the observation point at the current moment. Scholars at home and abroad have done a lot of research on clustering algorithms and data association algorithms.

At present, there are four types of clustering algorithms: the first type is partition-based clustering algorithm; the second type is density-based clustering algorithm; the third type is grid-based clustering algorithm; the fourth type is constraint-based clustering algorithm Clustering Algorithm. The K-Means clustering algorithm is a

Table 1.2 Main performance indicators of CSR-TS radar

Performance	Attributes
Number of lanes covered	4 ~ 5 Lane
Detection distance	30 ~ 200 m
Speed range	5 ~ 250 km/h
Speed accuracy	1 km/h
Traffic flow	The total traffic flow is greater than or equal to 98%, Single-lane traffic flow is greater than or equal to 95%
Lane occupancy	Single lane occupancy rate greater than or equal to 95%, The total lane occupancy is 98%
Congestion information judgment accuracy	90%
Recognition of wrong-way vehicles	95%
Lane-changing vehicle recognition accuracy	90%
Data update rate	$\leq$ 100 ms

typical partition-based clustering algorithm. The algorithm must determine a cluster center for each category in advance. It has a slow convergence speed, can only handle numerical data, and is sensitive to outliers. The algorithm has been improved. Pelleg et al. proposed the X-Means algorithm, which improved the convergence speed of the algorithm; Nguyen et al. proposed the K-Modes algorithm, which expanded the data types applicable to the algorithm; Bezdek proposed the Fuzzy C-Means (FCM) clustering algorithm, and the membership degree is used to represent the similarity between sample points, which is a soft clustering algorithm with good convergence.

The DBSCAN (Density-based Spatial Clustering of Applications with Noise) clustering algorithm is based on density division, which divides the regions with sufficient density into one category, and the clustering result is the largest set of points connected by density. Birant et al. improved the DBSCAN clustering algorithm and proposed the ST-DBSCAN (Spatial–Temporal DBSCAN) clustering algorithm, which can more accurately cluster data of different dimensions; Qian Pengjiang et al. proposed an adaptive GRC (Graph-based Relaxed Clustering) clustering algorithm; Mihael Ankers proposed the OPTICS (Ordering Points to Identify the Clustering Structure) clustering algorithm on the parameter order structure; Martin Ester proposed the DENCLUE (Density-based Clustering) clustering algorithm; Wang proposed the STING clustering algorithm.

Data association algorithms include single-objective association algorithms and multi-objective association algorithms. In 1971, Singer proposed the Nearest Neighbor Data Association (NNDA) algorithm, which works well when applied to single target tracking. The algorithm determines the observation point closest to the target prediction point as the associated point trace, and uses this observation point to continue the state update. In the case of less clutter and scattered targets, the execution efficiency of the algorithm is very high, but in dense multi-target scenes or near target scenes, there will be errors in association, such as low accuracy and weak anti-interference ability question. Some scholars have improved the nearest neighbor data association algorithm, and proposed the Global Nearest Neighbor Data Association (GNNDA) algorithm, which uses the minimum total statistical distance to find the associated point traces, and gradually formed a Bayesian-based "GNNDA" algorithm. "Soft Association" algorithm.

The Probability Data Association (PDA) algorithm is a suboptimal filtering algorithm proposed by Bar-Shalom and Jaffer. Because the PDA algorithm is applied under the assumption that there is only one target in the association area, the number of targets that can be processed when using this algorithm for target association is similar to that of the nearest neighbor data association algorithm. In order to solve the multi-objective data association problem, Bar-Shalom proposed the Joint Probability Data Association (JPDA) algorithm. The JPDA algorithm selects the most suitable event by calculating the probability of occurrence of all joint events, resulting in an exponential increase in the amount of computation with the increase of radar echo signals. Although the algorithm can solve the multi-objective data association problem well, the real-time performance is not good. In the research of data association algorithm, domestic experts and scholars have also made outstanding contributions. The more representative one is Professor Pan Quan of Northwestern Polytechnical University. Based on the JPDA algorithm, he proposed a generalized probabilistic data association (Generalized Probability Data Association, GPDA) algorithm. The algorithm combines more related event information, and the calculation amount only increases linearly with the increase of radar echo signals. Compared with the JPDA algorithm, it has higher accuracy and stronger real-time performance. Reid et al. proposed the Multiple Hypothesis Tracking (MHT) algorithm, which considers the multiple possibilities of observation points originating from existing targets, new targets and background clutter, and solves the target tracking problem through evaluation. Because the MHT algorithm considers almost all situations, its tracking accuracy is high and the real-time performance is not good.

Many scholars have made improvements and innovations on the basis of the above algorithms. For example, Musicki and Evans proposed the Joint Integrated Probability Data Association (JIPDA) algorithm; Song et al. proposed the Probabilistic Multiple Hypothesis Tracking (PMHT) algorithm; Turkmen et al. proposed a simplified joint probability data Association (Cheap Joint Integrated Probability Data Association, CJPDA) algorithm; Roecker et al. proposed Suboptimal Joint Integrated Probability Data Association (SJPDA) algorithm; Fitzgerald et al. proposed Exact Nearest Neighbor PDA (Exact Nearest Neighbor PDA, ENNPDA)) algorithm; Blom and Bloem proposed the Coupled Probability Data Association (CPDA) algorithm.

1.3 Radar Speed Measurement System

As a typical microwave electronic measuring instrument, the traditional traffic radar speed measurement equipment is mainly based on the Doppler speed measurement principle place.

The radar speed measurement system is an important measurement equipment in the field of intelligent transportation. If it is found that a target enters the radar irradiation area through real-time signal analysis, the signal processor inside the radar will process the data returned by the microwave front-end to calculate the speed of the target and other information. Pass it to the upper computer (industrial computer). At the same time, signals can be provided to cameras connected to the radar to record relevant information and pass all information to the relevant traffic authorities through fixed channels.

The radar speed measurement system is shown in Fig. 1.4, which mainly includes the radar signal processing system, the CCD camera control unit, the upper computer information processing system and the wireless transmission system, which are briefly described.

1. Radar signal processing system

   The effective identification and accurate tracking of the target vehicle by the radar speed measurement system is the focus of the research. The center frequency of the radar operation is 24.15 GHz. The radar antenna transmits the signal and is reflected back by the target vehicle. After the antenna receives it, it undergoes quadrature mixing and amplification to convert the analog signal. The digital signal is converted into a digital signal and transmitted to the digital signal processor. After the digital signal processor performs correlation operation and processing, it determines the information related to the target vehicle, including speed, direction, distance, etc. Determine whether to trigger the camera to capture the target vehicle and record information according to the specific working state.
2. CCD camera control unit

   The role of the camera is not only to capture, when the camera receives the capture signal from the radar, it immediately captures and receives the speed, distance and other information transmitted by the radar. At the same time, video and image processing are also performed to extract useful information such as the

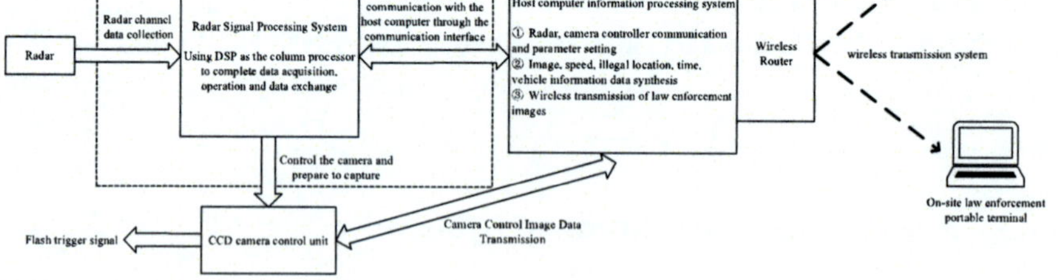

Fig. 1.4 Radar speed measurement system

license plate of the vehicle from the captured image to ensure that the information is complete and effective.

3. Host computer information processing system

   The main function of the upper computer information processing system is to receive and process the relevant information returned by the radar and the camera, and to classify and process the acquired information according to the work requirements and store it in the database. Through the wireless transmission system, the host computer can transmit relevant information to the traffic management department or traffic law enforcement department according to specific needs, which has the characteristics of real-time and effective.

4. Wireless transmission system

   The wireless transmission system is mainly composed of wireless modems or wireless routers. In order to transmit data in a short distance, the host computer is connected with the wireless router; if it needs to realize long-distance information transmission with the traffic management department, the host computer must be connected with the wireless modem, and the working methods such as dialing are adopted.

1.4 Research Status of Video Object Detection

The purpose of object detection is to obtain the location and class of one or more objects from an image. Traditional target detection algorithms rely more on feature operators designed subjectively by humans, such as HOG features and SIFT operators. But in most cases, feature extraction algorithms have certain limitations and are difficult to use for other detections. In 2008, Felzenszwalb proposed the DPM algorithm based on the idea that different components of the target can be combined into a spring model. The algorithm uses root filters and component filters to extract image features, and uses Latent Support Vector Machine (LSVM) Train the gradient model of the target and use the model as a template for target matching. In recent years, the rise of deep learning, especially the development of convolutional neural networks, has provided a new idea for feature extraction, that is, to avoid subjective feature design and use models to automatically complete feature extraction from data.

At present, target detection based on convolutional neural network can be divided into two categories: R-CNN series algorithms based on Region Proposal and YOLO algorithm based on Grid Partition. In 2013, Ross Girshick proposed the R-CNN algorithm, which abandoned the exhaustive search algorithm of the traditional sliding window, and adopted the idea of "clustering" to give 1000 to 2000 candidate regions that may be targets in advance, and then use the convolutional neural network to carry out the algorithm. Feature extraction, and classification and regression by SVM classifier and Bounding box regressor, respectively. The R-CNN algorithm introduced deep learning into the field of target detection for the first time, and greatly improved the detection rate. The detection rate on PASCAL VOC increased from 35.1 to 53.7%. In 2015, in order to solve the problem that the R-CNN algorithm is too slow, Ross Girshick further optimized the R-CNN algorithm and proposed the Fast R-CNN and Faster R-CNN algorithms. The Fast R-CNN algorithm draws on the idea of Spatial Pyramid Pooling of SPP-Net, cancels the image normalization process, solves the problem of information loss caused by image deformation, and makes the algorithm process more compact. The detection rate has been increased to about 67%, and the detection speed has also been greatly improved. With the same network size, compared to R-CNN, the training time of Fast R-CNN is shortened from 84 to 9.5 h, and the test time is shortened from 47 to 0.32 s. The biggest improvement of the Faster R-CNN algorithm is the proposed RPN (Region Proposal Network), which no longer uses a separate step to extract candidate regions, unifies all the steps of target detection into a deep learning framework, and realizes end-to-end detection. The target detection algorithm with the highest recognition rate.

Because the speed of the R-CNN series of algorithms still cannot meet the needs of real-time detection. In 2015, Joseph Redmon proposed the YOLO algorithm based on grid division. Different from the algorithm based on the candidate region, the YOLO algorithm regards target detection as a regression problem. The input image can directly obtain the position, category and confidence probability of all targets through a network inference, and can optimize the target detection performance end-to-end. Therefore, the detection speed of the YOLO algorithm is much faster than the R-CNN series of algorithms, up to 45 fps on the GPU of Titan X. However, the recognition rate of YOLO algorithm for target detection is low, and the detection effect of small targets is not very satisfactory, because the grid division leads to serious feature loss and lacks multi-scale regression. In 2015, Liu Wei proposed the SSD (Single Shot Multibox Detector) target detection algorithm, which uses the network layers with different receptive fields in the convolutional neural network to extract feature maps of different scales, and uses multi-scale feature maps. Improve small object detection problem.

1.5 Summary

This chapter introduces the development of the field of intelligent transportation, points out the defects of related technologies and proposes improvement plans, introduces the relevant characteristics of millimeter-wave radar and the concept of using millimeter-wave radar to build a multi-target tracking system. This chapter selects several representative traffic radars, and focuses on their characteristics and performance indicators. It also introduces the relevant radar signal processing technology and the important structure of the radar speed measurement system, as well as the video data processing widely used in traffic scene monitoring.

Chapter 2
Traffic Radar System

The radar transmits electromagnetic waves and processes the received echo signals to obtain valid signals. The emitted electromagnetic waves can be divided into two categories: pulsed and continuous waves. Radars that use them as emitted waves are called pulsed radars and continuous wave radars.

The former distance measurement is relatively simple, and the target distance is calculated by using the time difference between the transmitted pulse and the received echo signal. But in the traffic bayonet radar, it is generally required to measure the target within 100 m. In the case of such a short distance, the pulse width is required to be narrow so that the distance resolution can reach the usable range, which is difficult to achieve in engineering. In addition, the peak value of transmit power is also required to be large, which makes it difficult to realize multi-target monitoring.

The latter's transmit power does not change significantly over time, so it has better speed and ranging performance, and it is relatively easy to implement in hardware. Because the transmitted wave is continuous in time which means the duty cycle is 100%, the transmit power is much lower than the peak power of the pulse radar when the average power is the same. Therefore, continuous wave radar is mainly studied in the field of transportation. The commonly used continuous wave radar systems include unmodulated continuous wave (CW) system, linear frequency modulated continuous wave (LFMCW) system, frequency keying (FSK) system and CW-FMCW composite system.

2.1 CW System

The CW system mainly uses the Doppler effect. The Doppler effect refers to the change in the received frequency when there is relative radial motion between the source and the observer. The principle of the Doppler effect is shown in Fig. 2.1.

L. Cao et al., *Target Recognition and Tracking for Millimeter Wave Radar in Intelligent Transportation*, https://doi.org/10.1007/978-981-99-1533-0_2

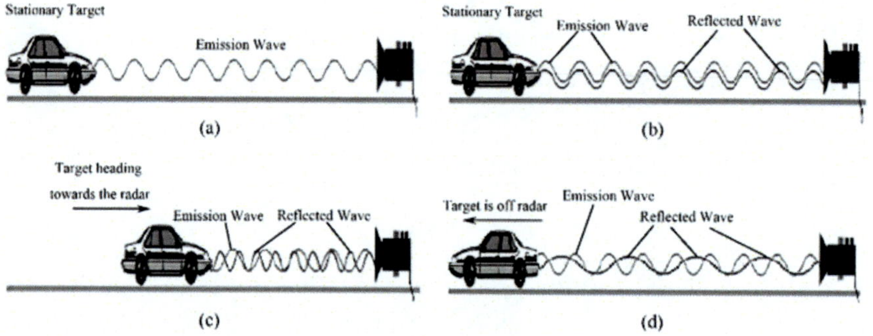

Fig. 2.1 Principle of Doppler effect

The continuous wave transmit signal is

$$s(t) = A\cos(\omega_0 t + \phi) \tag{2.1}$$

In the formula, ω_0 is the emission angular frequency, ϕ is the initial phase, and A is the amplitude.

The echo signal is

$$s_r(t) = Ks(t - t_r) = KA\cos[\omega_0(t - t_r) + \phi] \tag{2.2}$$

In the formula, $t_r = \frac{2R}{c}$ is the delay time of the echo signal to the transmitted signal; c represents the propagation speed of the electromagnetic wave in the air; R is the distance between the target to be measured and the radar antenna, and K is the signal attenuation coefficient.

If the measured target is stationary, the distance does not change. There is a constant phase difference ($\omega_0 t_r = 2\pi f_0 \frac{2R}{c} = 2R\frac{2\pi}{\lambda}$) between the echo signal and the transmitted signal. λ is the wavelength, $\lambda = \frac{c}{f_0}$; f_0 is the emission frequency of the electromagnetic wave.

If the measured target moves, the distance will change. If the target moves at a uniform speed, then at time t, the distance between the target and the radar antenna is

$$R(t) = R_0 - \upsilon_r t \tag{2.3}$$

In the formula, R_0 is the distance between the target and the radar antenna at $t = 0$; υ_r is the radial velocity of the target relative to the radar movement.

It can be known from Eq. (2.2) that the transmitted signal at time $t - t_r$ is received at time t. Because the propagation speed c of electromagnetic waves in the air is much greater than the relative running speed between the target and the radar, it is approximated as

$$t_r = \frac{2R(t)}{c} = \frac{2}{c}(R_0 - \upsilon_r t) \tag{2.4}$$

The phase difference between the transmitted signal and the echo signal is

$$\phi = -\omega_0 t_r = -\omega_0 \frac{2}{c}(R_0 - \upsilon_r t) = -2\pi\frac{2}{\lambda}(R_0 - \upsilon_r t) \tag{2.5}$$

φ is a function of time t, when the speed υ_r is constant, the frequency difference is

$$f_d = \frac{1}{2\pi}\frac{\mathrm{d}\phi}{\mathrm{d}t} = \frac{1}{2\pi} \times (2\pi\frac{2}{\lambda}\upsilon_r) = \frac{2}{\lambda}\upsilon_r \tag{2.6}$$

f_d is the Doppler frequency shift, which is inversely proportional to the radar wavelength and proportional to the relative running speed of the radar and the target. When the target and the radar are close to each other, the Doppler frequency shift is positive, and the frequency of the received signal is higher than the frequency of the transmitted signal. When the measured target and the radar are far away from each other, the Doppler shift is negative, and the frequency of the received signal is lower than the frequency of the transmitted signal.

When the vehicle drives towards the speed measuring radar, the electromagnetic wave encounters the echo signal reflected back by the vehicle and is compressed, and the frequency of the echo signal increases with the increase of the speed of the vehicle; on the contrary, when the motor vehicle drives away from the speed measuring radar, the electromagnetic wave is compressed. When encountering the echo signal reflected by the vehicle, it is amplified, and the frequency of the echo signal decreases with the increase of the speed of the vehicle. When the electromagnetic wave encounters the target to be measured and reflects back, the relationship between the frequency of the echo signal and the frequency of the electromagnetic wave emitted by the radar is:

$$f_0' = f_0 + \frac{2\upsilon_r}{c} f_0 \tag{2.7}$$

In the formula, f_0' is the frequency of the echo signal; f_0 is the frequency of the electromagnetic wave emitted by the radar; υ_r is the radial velocity component of the moving target.

It can be seen from Eq. (2.7) that the frequency of the echo signal received by the radar consists of two parts, one part is the frequency of the electromagnetic wave emitted by the radar, and the other part is the frequency offset of the echo signal caused by the movement of the vehicle relative to the radar, when the vehicle is moving towards the radar, υ_r is a positive number, and when the vehicle is moving away from the radar, υ_r is a negative number, the formula (2.7) can be expressed as

$$f_d = f_0' - f_0 = \frac{2\upsilon_r f_0}{c} \tag{2.8}$$

From the formula (2.8), the radial velocity of the vehicle is obtained as

$$v_r = \frac{c}{2f_0} f_d \tag{2.9}$$

It can be known from Eq. (2.9) that the radial velocity of the vehicle is proportional to the Doppler frequency shift, so it can be calculated by measuring.

2.2 LFMCW System

The FMCW system uses high-frequency signals such as triangular waves or saw tooth waves as modulation signals, and performs frequency modulation on the transmitted signal to form a frequency-modulated continuous wave. For a stationary target, it mixes the echo signal and the transmitted signal to obtain a difference frequency signal, and after processing, the distance information of the target is obtained; for a moving target, it obtains the speed information of the target according to the simultaneous Doppler frequency shift.

Under the FMCW system, the emitted wave of the radar is a constant amplitude wave, and the frequency changes periodically over time. In each cycle, the frequency is linearly related to time, and the distance and speed of the target are obtained by detecting the frequency of the echo signal. In the LFMCW system, the frequency of the carrier signal has a linear relationship with time. Within a certain interval, the frequency of the carrier frequency signal and the voltage of the modulation signal have a linear relationship, and the curves of the two changing with time have similar shapes. The following takes the triangular wave as an example to illustrate the principle of speed measurement and ranging of the LFMCW system. The relationship among triangular wave modulation voltage, frequency and time is shown in Fig. 2.2.

The expression for the voltage is

$$\begin{cases} \Delta V = V_{\max} - V_{\min} \\ V_{\text{mod}} = V_{\min} + \frac{\Delta V}{T}(t - 2nT), \quad 2nT \le t \le (2n+1)T \\ V_{\text{mod}} = V_{\min} + \frac{\Delta V}{T}[(t - 2n + 1)T], \quad (2n+1)T < t \le (2n+2)T \end{cases} \tag{2.10}$$

The frequency expression is

$$\begin{cases} f_{\max} = f_0 + kV_{\max} \\ f_{\min} = f_0 + kV_{\min} \\ \Delta f = f_{\max} - f_{\min} \\ f(t) = f_{\min} + \frac{\Delta f}{T}(t - 2nT), \quad 2nT \le t \le (2n+1)T \\ f(t) = f_{\max} + \frac{\Delta f}{T}[t - (2n+1)T], \quad (2n+1)T < t \le (2n+2)T \end{cases} \tag{2.11}$$

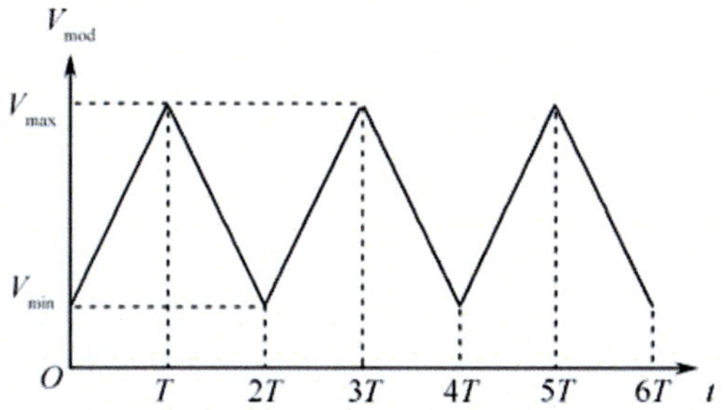

(a) Relationship between triangular wave modulation voltage and time

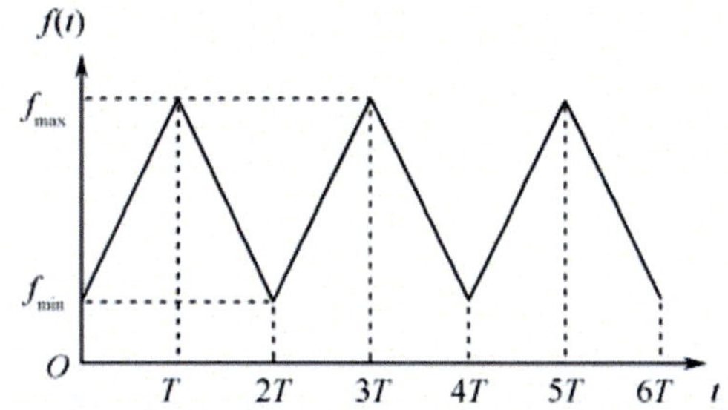

(b) The relationship between triangular wave modulation frequency and time

Fig. 2.2 The relationship between triangular wave modulation voltage, frequency and time

Figures 2.3 and 2.4 show the intermediate frequencies of stationary and moving targets respectively.

It can be seen from Figs. 2.3 and 2.4 that when the time is $[T + t_d, 2T]$, $f_m(t)$ is a negative number. Compared with the positive frequency, the negative frequency only indicates the speed of the phase angle changing in the opposite direction around the circumference. In fact, we can only read the absolute value of the frequency from the instrument, which represents how fast the phase changes. During the time period, the IF frequency changes linearly, and in the rest of the time period, the IF frequency remains unchanged. The intermediate frequency of the triangular wave modulation is

Fig. 2.3 IF frequency of stationary target

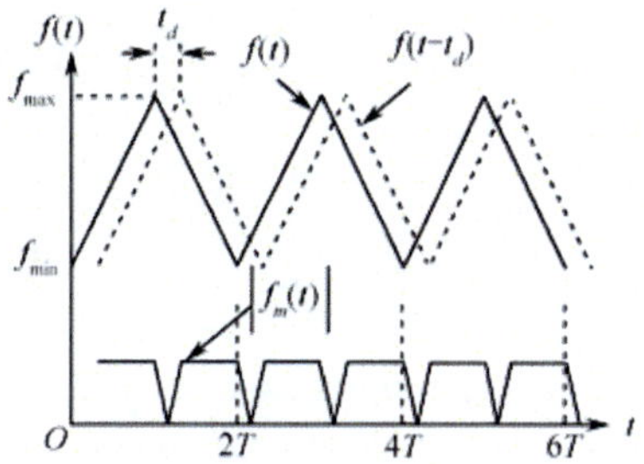

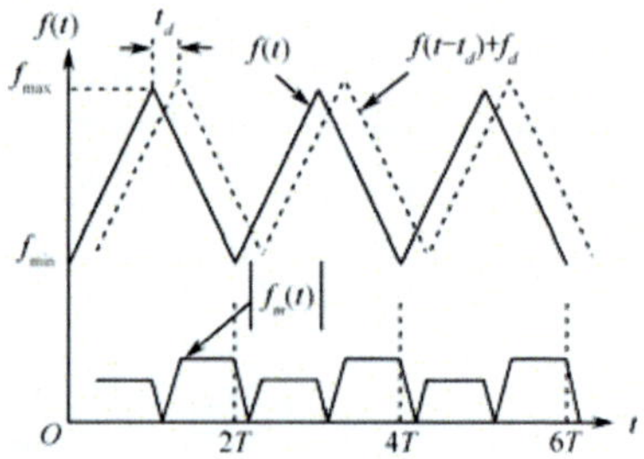

Fig. 2.4 IF frequency of moving target

$$\begin{cases} |f_{m1}(t)| = \frac{\Delta f}{t}\frac{2R}{c} - \frac{2v}{\lambda}, & 2nT + t_d \le t \le (2n+1)T \\ |f_{m2}(t)| = \frac{\Delta f}{T}\frac{2R}{c} - \frac{2v}{\lambda}, & (2n+1)T + t_d \le t \le (2n+2)T \end{cases} \tag{2.12}$$

From formula (2.12) we can get

$$\begin{cases} R = \frac{cT|f_{i1}(t_1) + f_{i2}(t_2)|}{4\Delta f} \\ v = \frac{\lambda|f_{i1}(t_1) + f_{i2}(t_2)|}{2} = \frac{c|f_{i1}(t_1) + f_{i2}(t_2)|}{2f_0} \end{cases} \tag{2.13}$$

In the formula, t_1 and t_2 represent two time periods respectively, $f_{i1}(t_1)$ and $f_{i2}(t_2)$ represent the intermediate frequency corresponding to the time periods t_1 and t_2 respectively.

The propagation speed c of the electromagnetic wave in the air, the period 2 T, the sweep frequency bandwidth Δf, and the center frequency f_0 are known. As long as we obtain the intermediate frequency $f_{i1}(t)$ and $f_{i2}(t)$ the corresponding time period, we can obtain the distance and speed of the target. The velocity of a stationary target is zero and there is no Doppler shift. The disadvantage of the LFMCW system is that it is not easy to obtain a good chirp degree, which leads to the deterioration of the range resolution, and needs to be combined with other systems to correctly distinguish multiple moving targets. If the two equations in Eq. (2.12) are the same, then we have

$$\begin{cases} f_i(t) = \frac{\Delta f}{T}\frac{2R}{c} \\ \Delta f = \frac{Tf_i(t)}{t_d} \\ R = \frac{cTf_i(t)}{2\Delta f} \end{cases} \tag{2.14}$$

When multiple stationary targets with different distances are found in the radar irradiation area, because there is no Doppler frequency shift, and each frequency of the signal corresponds to a distance, the radar can distinguish different targets. When there are multiple moving targets with different distances ahead, a set of frequencies twice as many as the targets will be generated in the frequency spectrum of the intermediate frequency signal. Therefore, for a target, it is impossible to determine the unique frequency associated with it and to distinguish the target, even false targets that do not exist. In this case, it is necessary to combine other modulation methods to correctly identify the target, such as triangular wave modulation with the same amplitude but different periods, and triangular wave modulation with the same period and different amplitudes.

2.3 FSK System

Under the FSK system, the radar transmits continuous waves and receives echo signals. The modulated signal is a square wave signal that is a signal with frequencies f_1, f_2. Figure 2.5 shows the working principle of the FSK system.

The transmitted signal is

$$V_t(t) = \begin{cases} A_{t1}\cos[\varphi_1(t)] = A_{t1}\cos(2\pi f_1 t + \varphi_0), & 2nT \le t \le (2n+1)T \\ A_{t2}\cos[\varphi_2(t)] = A_{t2}\cos\left(2\pi f_2 t + \varphi_0'\right), & (2n+1)T < t \le (2n+2)T \end{cases} \tag{2.15}$$

The echo signal is

$$\begin{aligned} V_r(t) &= \begin{cases} A_r\cos[2\pi f_1(t - t_d) + \varphi_0], & 2nT \le t \le (2n+1)T \\ A_r\cos\left[2\pi f_2\left(t - t_d'\right) + \varphi_0'\right], & (2n+1)T < t \le (2n+2)T \end{cases} \\ &= \begin{cases} A_r\cos[\varphi_1(t) - 2\pi f_1 t_d], & 2nT \le t \le (2n+1)T \\ A_r\cos\left[\varphi_2(t) - 2\pi f_2 t_d'\right], & (2n+1)T < t \le (2n+2)T \end{cases} \end{aligned} \tag{2.16}$$

After mixing and demodulation, the intermediate frequency signal is obtained as

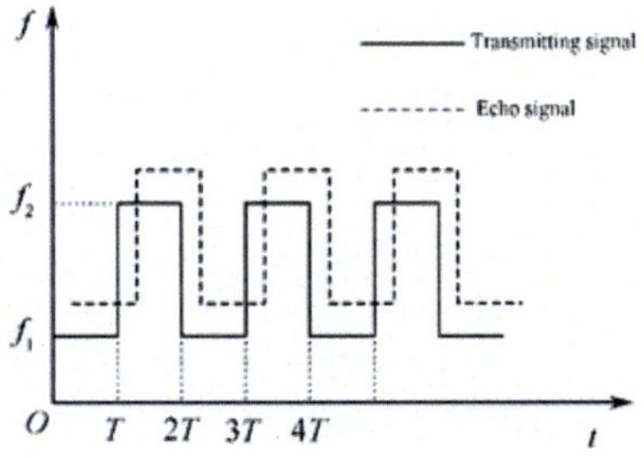

Fig. 2.5 The working principle of the FSK system

$$
\begin{aligned}
V_i(t) &= \begin{cases} A_i \cos[\varphi_{i1}(t)], & 2nT + t_d \le t \le (2n+1)T \\ A_i \cos[\varphi_{i2}(t)], & (2n+1)T + t_d \le t \le (2n+2)T \end{cases} \\
&= \begin{cases} A_i \cos(2\pi f_1 t_d), & 2nT + t_d \le t \le (2n+1)T \\ A_i \cos\left(2\pi f_2 t_d'\right), & (2n+1)T + t_d \le t \le (2n+2)T \end{cases} \\
&= \begin{cases} A_i \cos\left(\dfrac{4\pi f_1}{c} R\right), & 2nT + t_d \le t \le (2n+1)T \\ A_i \cos\left(\dfrac{4\pi f_2}{c} R'\right), & (2n+1)T + t_d \le t \le (2n+2)T \end{cases}
\end{aligned}
\tag{2.17}
$$

$$
R = R_0 - vt \tag{2.18}
$$

$$
R' = R_0' - v't \tag{2.19}
$$

In the formula, R_0 is the distance between the radar and the target when the signal $V_1(t)$ with frequency f_1 reached the target, v is the radial velocity of the target; R_0' is the distance between the radar and the target when the signal $V_2(t)$ with frequency f_2 reached the target, v' is the radial velocity of the target.

The frequency of the intermediate frequency signal is

$$
f_i = \begin{cases} \frac{1}{2\pi}\frac{\mathrm{d}\varphi_{i1}}{\mathrm{d}t} = \frac{1}{2\pi}\frac{\mathrm{d}}{\mathrm{d}t}\left(\frac{4\pi f_1}{c} R\right) = \frac{2f_1}{c}\frac{\mathrm{d}R}{\mathrm{d}t} \\ \quad = \frac{2f_1}{c} v = \frac{2}{\lambda_1} v = f_{d1}, \quad 2nT + t_d \le t \le (2n+1)T \\ \frac{1}{2\pi}\frac{\mathrm{d}\varphi_{i2}}{\mathrm{d}t} = \frac{1}{2\pi}\frac{\mathrm{d}}{\mathrm{d}t}\left(\frac{4\pi f_2}{c} R'\right) = \frac{2f_2}{c}\frac{\mathrm{d}R'}{\mathrm{d}t} \\ \quad = \frac{2f_2}{c} v' = \frac{2}{\lambda_2} v' = f_{d2}, \quad (2n+1)T + t_d \le t \le (2n+2)T \end{cases}
\tag{2.20}
$$

In the formula f_{d1} and f_{d2} are the Doppler shifts of the two echo signals, respectively.

The center frequency of the radar operation is 24.15 GHz, and the swept frequency bandwidth is $\Delta f = |f_1 - f_2| = 1.5$ MHz which is very small compared to the center frequency of the signal. In the calculation of Doppler frequency shift, if the frequencies of the two transmitted signals are used f_1 as substitutes, the relative error is $\delta = \Delta f / f_1 = 0.006\%$ which is almost indistinguishable for the radar system used, so it can be considered that the two signals have the same Doppler frequency shift. The speed of the target can be obtained from the formula (2.3). When there are multiple moving targets, the targets with different speeds are also different. The FSK system radar can distinguish the targets in the spectrum according to this characteristic.

It can be seen from Eq. (2.17) that the intermediate frequency signal of each moving target can be demodulated into two signals with different phases but almost the same frequency. By analyzing their phases, the target distance can be obtained. The relationship between the phase difference and the frequency difference of the two signals is

$$\Delta\varphi = \varphi_1 - \varphi_2 = \frac{4\pi R(f_1 - f_2)}{c} = \frac{4\pi R\Delta f}{c} \tag{2.21}$$

The target distance can be obtained as

$$R = \frac{c\Delta\varphi}{4\pi\Delta f}, \ \Delta\varphi \in (-\pi, \pi) \tag{2.22}$$

Since the frequency of the radar carrier frequency signal f_1, f_2 is known, the swept frequency bandwidth Δf can be obtained. As long as the phase difference $\Delta\varphi$ of the two signals is calculated, the target distance can be obtained.

The FSK system radar transmits a dual-frequency continuous wave, and the two frequencies of the signal appear alternately. The Doppler frequency shift is used to measure the speed, and the target distance is obtained through the phase difference of the two carrier frequencies in the receiver. The FSK system radar is easier to implement, has better anti-noise and anti-attenuation performance, and can measure the target speed, distance and multiple targets with different speeds. However, it distinguishes multiple targets by the Doppler frequency shift of the target, so it can only detect moving targets, not stationary targets. This feature is just an advantage when applied to roads, which can exclude roads, manhole covers and surrounding railings, etc. Static clutter interference generated by stationary objects improves the signal-to-noise ratio of the target.

2.4 CW-FMCW Composite System

At present, most of the commonly used speed measuring radars use the CW system, which can only detect the speed of the target. The radar trigger is only judged by the strength of the target echo signal, and the trigger is triggered when the strength of the echo signal reaches the set value. However, in the actual road traffic environment, the size, structure, shape and material of vehicles are different, resulting in unstable signal strength. The trigger distance of the vehicle is close or far, and the effective information of the vehicle cannot be obtained from the captured images, and the effective capture rate can only be it reaches 90%, which seriously affects the supervision work of the traffic management department and cannot meet the increasing application needs of the intelligent transportation industry.

Although FMCW radar has excellent range resolution and can obtain range and speed information at the same time, there is a coupling effect between range frequency shift and Doppler frequency shift, resulting in low accuracy of speed and range. Moreover, when there is a stationary object in its irradiation range, that is, there is a fixed strong interference, it will seriously interfere with the judgment of the radar, and false targets may appear. If there are multiple targets at the same time, abnormal phenomena such as ultra-high speed or ultra-low speed may also occur due to the mismatch of I and Q channels. The coupling between the speed and distance

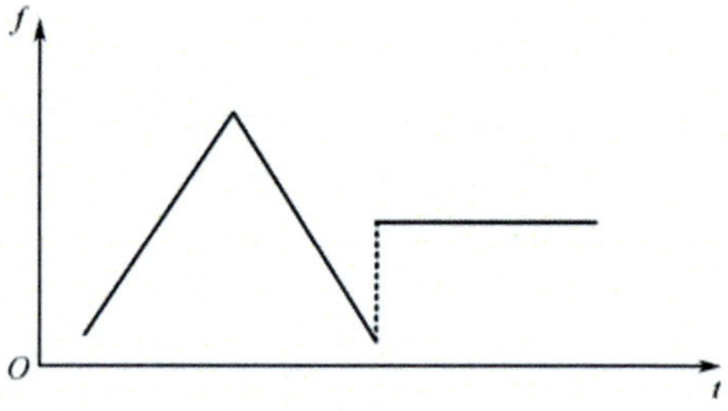

Fig. 2.6 CW-FMCW composite radar transmit signal waveform

of multiple targets will even cause the spectrum aliasing of each target, so that the target cannot be distinguished, resulting in a complete error in the speed and distance detection.

In 2001, Professor H. Rohling of Germany proposed the FSK-LFMCW composite system, which not only improved the overall performance of the radar, but also reduced the design difficulty of the hardware circuit. Inspired by this, the researchers deeply studied the working principles of the two systems and combined with the technical level of existing projects, and proposed a CW-FMCW complex to solve the problem that the common CW system radars on the market currently cannot perform high-precision positioning of targets. The problem of low effective capture rate.

Figure 2.6 shows the waveform of the CW-FMCW composite radar transmit signal. In Fig. 2.6, the part where the frequency changes linearly with the triangular wave modulation corresponds to the FMCW system, which is used to detect the speed and distance of the target; the part where the frequency does not change with time corresponds to the CW system with a fixed transmission frequency, which is used to accurately detect the target's speed and distance speed.

Under the CW-FMCW composite system, the signals of the two systems do not work simultaneously. Therefore, it is necessary to improve the hardware circuit (add a switch circuit in the hardware circuit), and realize the time-sharing switching of the modulation waveform through the STM32F051 single-chip microcomputer, so that the two systems can work in an orderly interleaved time, that is, the time-sharing multiplexing of the two systems is realized., in order to ensure the real-time and validity of measuring moving target information.

The CW-FMCW composite system retains all the measurement performances of the two systems, which not only solves the problem of ambiguous speed measurement in the FMCW system, but also makes up for the inability of the CW system to measure the target distance.

In the DSP algorithm, the CW-FMCW composite system obtains accurate speed information through the CW system, and uses it to replace the speed under the FMCW system according to certain rules, thereby realizing the decoupling of distance and speed under the FMCW system. It not only improves the ranging accuracy, but also realizes the high-precision positioning of the target, and does not affect the speed measurement accuracy of the radar. By adjusting hardware-related parameters and performing fusion processing of DSP algorithms, the radar can achieve high-precision positioning of the target and improve the effective capture rate of the radar.

2.5 Summary

This chapter introduces the speed and ranging principles of several commonly used traffic radar systems, and compares the advantages and disadvantages of each system. The CW system can only measure the speed of the target; the FMCW system can obtain the distance and speed of the target at the same time, but its measurement of the speed is not ideal. When there are multiple moving targets, the targets cannot be distinguished, and even false targets can be obtained; Although the composite system can obtain more accurate distance and speed, it has higher requirements for hardware; the FSK system is easier to implement, and has better anti-noise and anti-attenuation performance. Multiple targets, unable to detect stationary targets, can eliminate stationary clutter interference generated by stationary objects such as roads and surrounding railings, and improve the signal-to-noise ratio.

Chapter 3
Microwave Velocity Radar Signal Processing Algorithm

The electromagnetic wave signal emitted by the microwave speed measuring radar will be subject to various interferences when it propagates in the air, including the internal noise of the system, complex road conditions and environmental interference. When the distance between the target to be measured and the radar is far away, the external environment will interfere with the signal greatly, resulting in the inability to obtain an effective signal. Therefore, before the echo signal is converted from the time domain to the frequency domain and processed accordingly, the interference needs to be removed.

The traditional radar signal denoising method mainly performs windowing filtering on the time domain signal, including FIR filtering, median filtering and IIR filtering. For the filtering to be effective, the frequency band of the interfering signal and the frequency band of the actual signal should not overlap or the overlap part should be very small. However, the echo signal of the microwave speed measuring radar and the interference signal are aliased together. When the frequency of the target signal is low, this aliasing phenomenon is particularly prominent, and the frequency of the echo signal is related to the specific position of the target to be measured, so its frequency spectrum. The distribution is also constantly changing. The traditional windowing and denoising method can only suppress the interference signal in a fixed frequency band, and cannot adaptively adjust the filtering parameters according to the frequency of the echo signal. Therefore, the traditional windowing and denoising method cannot effectively remove the noise interference in the echo signal.

3.1 Denoising Algorithm

In order to eliminate noise interference in radar echo signals and extract high-quality target signals for processing, two denoising algorithms are introduced below.

L. Cao et al., *Target Recognition and Tracking for Millimeter Wave Radar in Intelligent Transportation*, https://doi.org/10.1007/978-981-99-1533-0_3

3.1.1 EMD-Based Denoising Algorithm

Empirical Mode Decomposition (EMD) is a signal analysis method proposed for time-domain signals, which decomposes the signal into several Intrinsic Mode Function (IMF) components, which represent the decomposition of the signal's partial frequency features. From the perspective of filtering, the empirical mode decomposition process is equivalent to passing the signal through different filters to obtain all the frequency characteristics of the signal, and recombining the non-noise signal by screening the noise signal to achieve the purpose of denoising.

Basic Principles and Filtering Characteristics of EMD

EMD believes that each split signal is composed of multiple signals with different frequencies, and the IMF is obtained by combining these signals. IMF must satisfy two conditions: ① On the complete signal sequence, the number of extreme points and the number of zero-crossing points are equal or different by one; ② At any point, the mean value of the upper and lower envelopes determined by the maximum and minimum values of the signal is 0.

The EMD decomposition process is shown in Fig. 3.1.

The signal can be split into n IMF components and 1 remainder, which is

$$x(n) = \sum_{i=1}^{n} C_i(n) + r(n) \tag{3.1}$$

In the formula, the IMF component contains different frequency components of the signal, and its frequency is from high to low, which represents the central trend. The empirical mode decomposition method also has the characteristics of completeness, orthogonality and adaptability.

The time domain signal waveform is shown in Fig. 3.2. The decomposition result is shown in Fig. 3.3.

It can be seen from Fig. 3.3 that 10 IMF components are obtained by empirical mode decomposition, of which the IMF components with small order correspond to the high-frequency components of the signal, mainly including the sharp parts of the signal and noise; The IMF component with a large order corresponds to the low-frequency component of the signal, which mainly includes the useful signal and a small amount of low-frequency noise components.

The traditional denoising method directly filters out the smaller-order part of the IMF component (the high-frequency component of the signal), and the effect of this method is general and it is possible to filter out the useful signal. EMD decomposes the signal into multiple IMF components with frequencies from high to low, and the frequency relationship is: The first IMF component contains the highest instantaneous frequency component, and the instantaneous frequency of the i-th IMF component is almost twice the instantaneous frequency of the (i + 1)-th IMF component, which reflects the multi-scale adaptive filtering characteristics. Compared with the original signal, each IMF component is a stable narrow-band

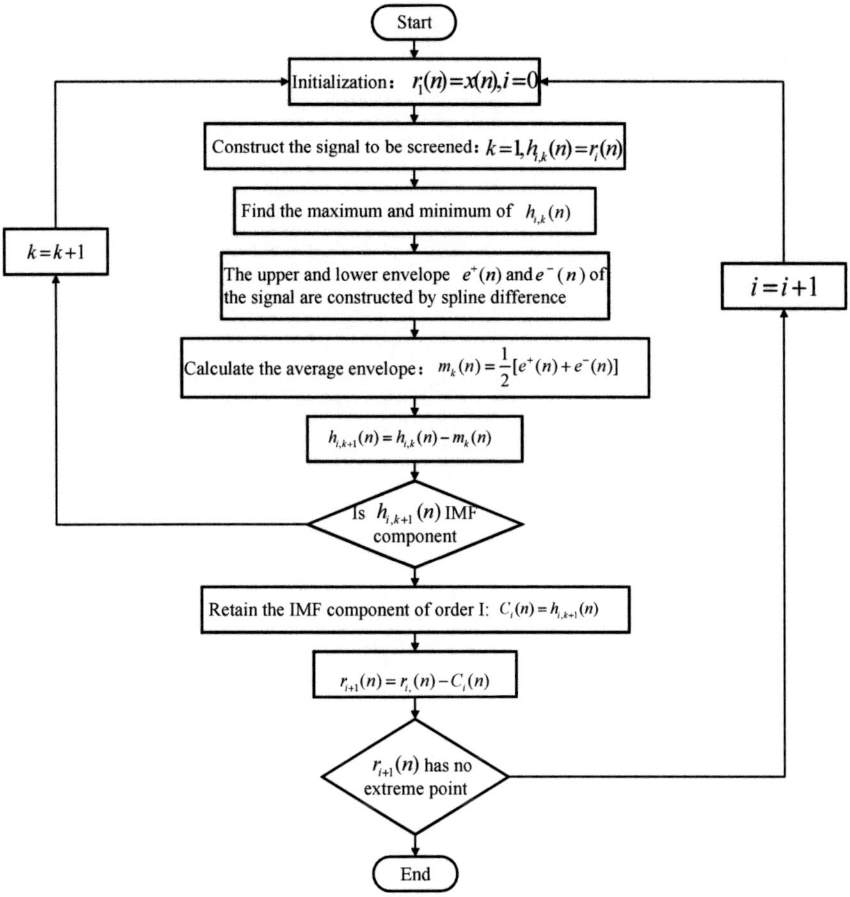

Fig. 3.1 EMD decomposition process

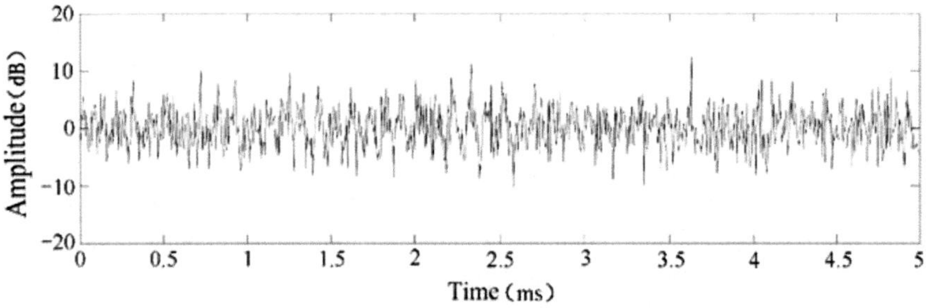

Fig. 3.2 Time domain signal waveform

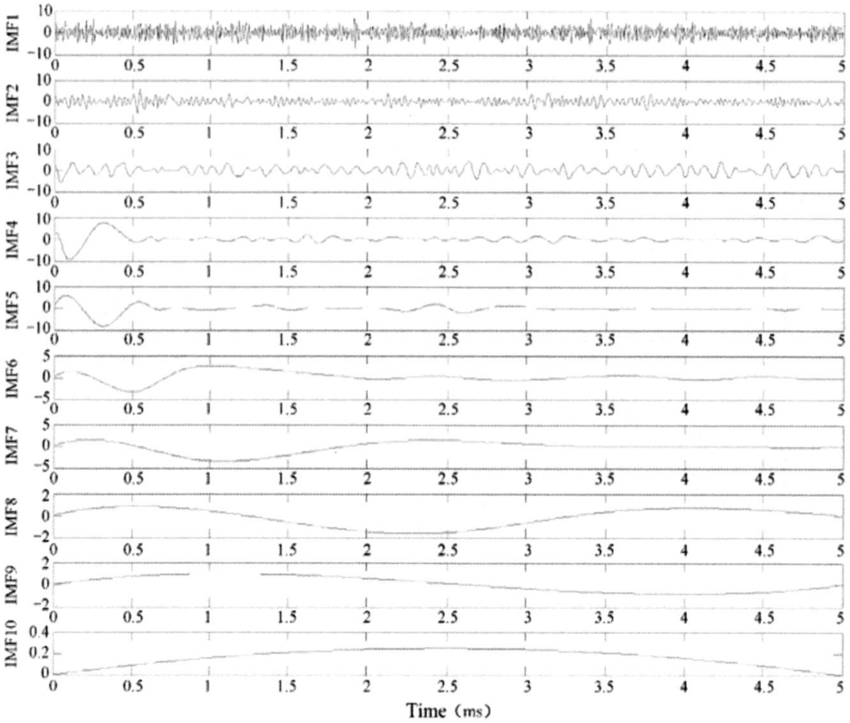

Fig. 3.3 Decomposition result

signal. According to the characteristics of the signal, some IMF components can be combined purposefully to highlight the characteristics of the original signal in a specific frequency range, so as to construct a new filter device. Therefore, the idea of denoising the difference frequency signal is to obtain multiple IMF components through EMD, and reconstruct the IMF components that contribute more to the useful signal, so as to reduce the influence of interference factors and improve the signal-to-noise ratio.

EMD Denoising Algorithm Based on Energy Characteristics of Autocorrelation Function

Boudella et al. studied an EMD denoising algorithm based on the continuous mean square error criterion, taking the global minimum point of the energy change curve of each IMF component as the boundary point between the dominant mode of the useful signal and the dominant mode of the noise. The algorithm improves the EMD denoising ability to a certain extent. When the signal-to-noise ratio is high, its effect is better; when the signal-to-noise ratio is low, its performance is not very stable, and even the global minimum value of the IMF component may not be found accurately, making the denoising algorithm completely meaningless.

Wang Ting et al. studied an EMD denoising algorithm based on the characteristics of the autocorrelation function. The algorithm uses the distribution characteristics of the autocorrelation function of the useful signal and the noise signal to determine the cutoff order of the adaptive filter in the EMD denoising algorithm. The denoising effect of the algorithm is good, and it is not limited by the signal-to-noise ratio. However, the algorithm determines the cutoff order by observing the distribution characteristics of the autocorrelation function of each IMF component, which is not direct enough. The correlation function is an average measure in the time-domain characteristics of a signal, which represents the similarity of different signals at different times. The time-averaged autocorrelation function is:

At different times. The time-averaged autocorrelation function is

$$R_{xy}(m) = \frac{1}{N - |m|} \sum_{n=0}^{N-1} x_N(n) y_N(n + m) \quad (3.2)$$

In the formula, m represents the time delay of the two signals. In order to ensure that the length of the autocorrelation function sequence is consistent with the length of the signal $x(n)$, the delay m should satisfy

$$-\frac{N}{2} - 1 < m < \frac{N}{2} \quad (3.3)$$

The sequence energy is

$$E = \sum_{n=0}^{N-1} x^2(n) \quad (3.4)$$

According to formula (3.2), the autocorrelation function of random noise and the autocorrelation function of general signal containing two sinusoidal components are calculated respectively. The time domain distribution of random noise and its autocorrelation function is shown in Fig. 3.4. The correlation of random noise is weak and the randomness is strong, which causes its autocorrelation function to achieve the maximum value at the zero point, and the autocorrelation result at other points is very small. Taking a sinusoidal signal as an example, the time domain distribution of a general signal and its autocorrelation function is shown in Fig. 3.5. The autocorrelation function does not decay rapidly to a small value at the zero point, but varies with the time difference. If the general signal is a periodic signal, its autocorrelation function is also a periodic signal, and its variation law is obviously very different from that of the autocorrelation function of random noise.

The energy comparison between random noise and its autocorrelation function and general signal and its autocorrelation function is shown in Table 3.1. Under the condition that the energy of random noise is not much different from that of general signal, the energy of the autocorrelation function of random noise is much lower than the energy of the autocorrelation function of general signal.

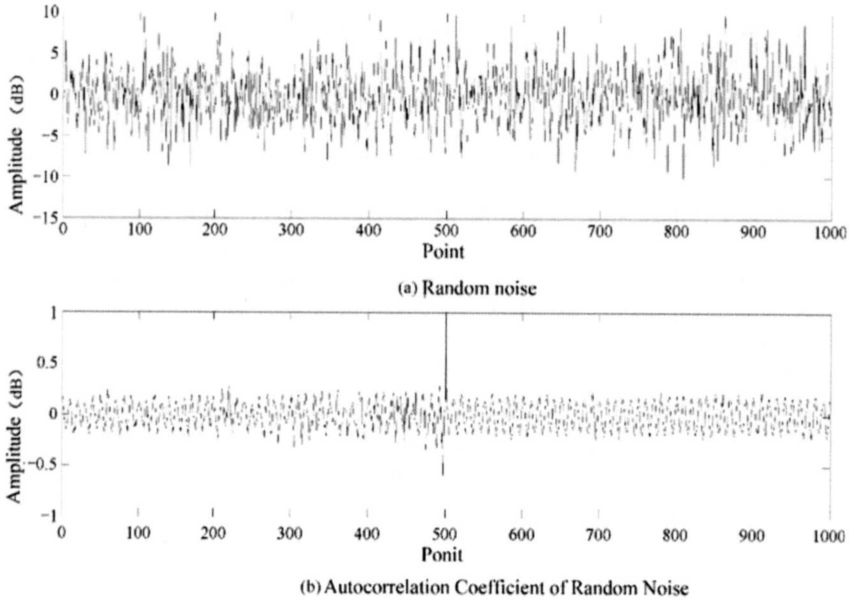

Fig. 3.4 Time domain distribution of random noise and its autocorrelation function

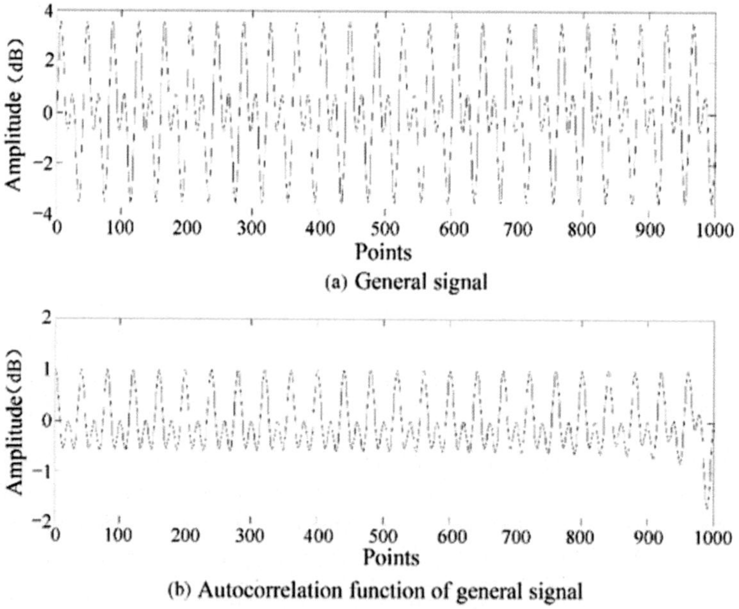

Fig. 3.5 Time domain distribution of general signals and their autocorrelation functions

Table 3.1 Energy comparison between random noise and its autocorrelation function and general signal and its autocorrelation function

	Self-energy (10^{-6} W)	Energy of autocorrelation function (10^{-6} W)
Random noise	499.50	0.52
General signal	499.69	125.40

Wu and Huang et al. found that white noise has certain statistical properties under the EMD algorithm. The IMF components obtained after the empirical mode decomposition of white noise all satisfy the normal distribution. Although the IMF components obtained after the empirical mode decomposition of Gaussian white noise are no longer white noise in the actual sense, their statistics are retained and its autocorrelation function is the largest at the zero point, and rapidly decays to zero at other points, which has the same distribution characteristics as the autocorrelation function of the original Gaussian white noise. The following is the empirical mode decomposition of the difference frequency signal containing Gaussian white noise, and the autocorrelation function of each IMF component is obtained as shown in Fig. 3.6. As can be seen from Fig. 3.6, the first three orders of the IMF component mainly contain the high frequency part of random noise, the fourth order mainly contains the useful signal, and the last six orders mainly contain the low frequency part of random noise.

The energy change curve of the autocorrelation function of each IMF component is shown in Fig. 3.7. As can be seen from Fig. 3.7, the energy of the autocorrelation

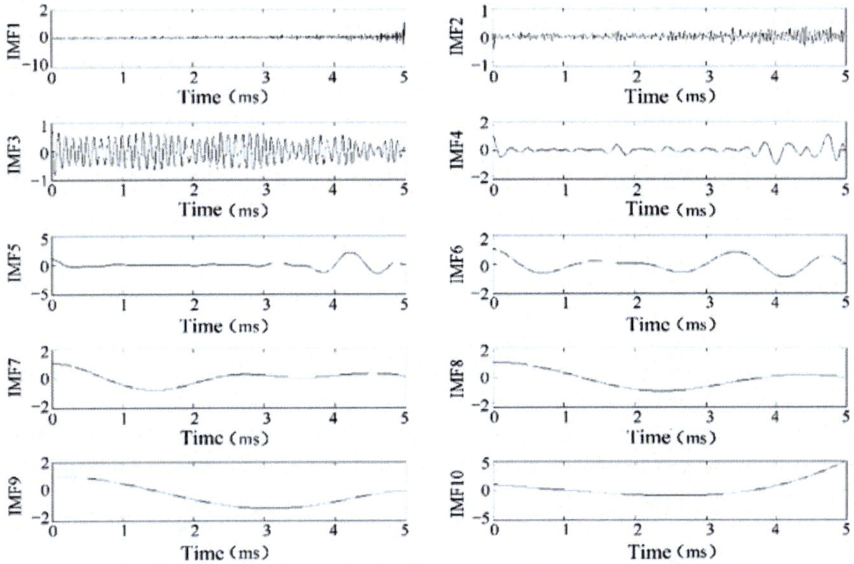

Fig. 3.6 Autocorrelation function of each IMF component

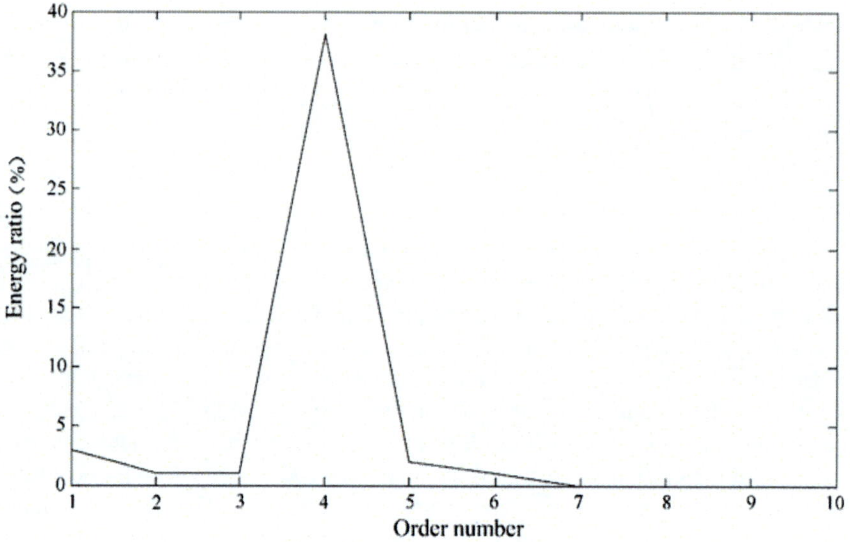

Fig. 3.7 Energy change curve of autocorrelation function of each IMF component

function of the fourth-order IMF component is the largest, which is the same as the result in Fig. 3.6, and the fourth-order mainly contains useful signals. Through Fig. 3.7, the demarcation point between the signal mode and the noise mode can be found more directly and accurately. The boundary point between the noise mode and the signal mode is the position i_m of the maximum value in the energy change curve of the autocorrelation function of each IMF component.

Because the first-order IMF component contains the most frequency components, its random noise energy is the largest, and the second-order IMF component contains half of the frequency components of the first-order. If the second-order IMF component is also dominated by random noise, then its energy will be greatly reduced. Therefore, when the signal-to-noise ratio is low, it may happen that the energy of the autocorrelation function of the first-order IMF component is larger than the energy of the IMF component that mainly contains the useful signal. When the signal-to-noise ratio is low, the energy change curve of the autocorrelation function of each IMF component is shown in Fig. 3.8. In this case, the IMF components that mainly contain useful signals should be determined according to the maximum position of the energy change curve of the autocorrelation function of the IMF components from the second order to the last order.

The time domain signal is an intermediate frequency signal, and a band-pass filter should be used to reconstruct the signal. Because the $i_m - 1$ th and $i_m + 1$ th order IMF components may contain the useful signal of the time domain signal, the signal should be reconstructed using the $i_m - 1$ th, i_m th and $i_m + 1$ th order IMF components. Among them, the $i_m - 1$ IMF component is dominated by noise, which should be denoised by soft threshold

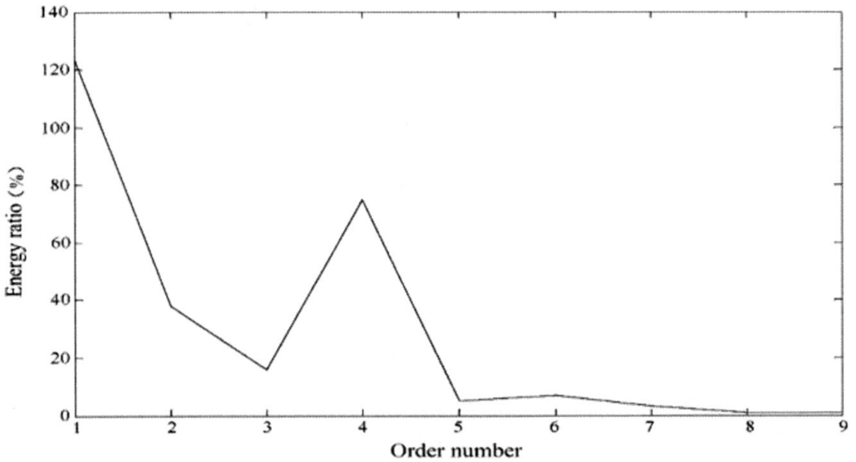

Fig. 3.8 The energy change curve of the autocorrelation function of each IMF component in the case of low signal-to-noise ratio

$$C'_{i_{m-1}}(n)\begin{cases} \operatorname{sgn}\left[C_{i_{m-1}}(n)\right]\left[\left|C_{i_{m-1}}(n)\right| - \lambda\right], & \left|C_{i_{m-1}}(n)\right| > \lambda \\ 0 & , \left|C_{i_{m-1}}(n)\right| < \lambda \end{cases} \tag{3.5}$$

In the formula, λ is the threshold of the $i_m - 1$ th order IMF component, and the calculation formula is

$$\lambda = \frac{\text{media}\left[\text{abs}(C_{i_m} - 1)\right]}{0.66745}\sqrt{2\ln N} \tag{3.6}$$

In the formula, media() is the function of taking the absolute median of a vector.

The calculation steps of the EMD denoising algorithm based on the energy characteristics of the autocorrelation function are as follows.

(1) Perform eigenmode decomposition on the input signal to obtain N eigenmode components.
(2) Calculate the unbiased autocorrelation function and the energy of the autocorrelation function of each IMF component respectively.
(3) According to the energy change curve of the autocorrelation function of each IMF component, start from the second order to find the position corresponding to its maximum value, and further find the order of the eigenmode component corresponding to the useful signal.
(4) Perform soft threshold denoising on the $i_m - 1$ th IMF components.
(5) Reconstructed signal $s(n) = C_{i_{m-1}}(n) + C_{i_m}(n) + C_{i_{m+1}}(n)$

Simulation and Analysis

Add the sine signal $s(n)$ to the randomly distributed Gaussian white noise to get $y(n)$, set the signal-to-noise ratio to SNR $= -5$ dB, and use the EMD denoising algorithm based on the energy characteristics of the autocorrelation function, using the signal-to-noise ratio (SNR) and root mean square error (RMSE) to evaluate the denoising effect.

The signal-to-noise ratio of the output signal is

$$\mathrm{SNR} = 10\lg \frac{\sum_{n=0}^{N-1} s^2(n)}{\sum_{n=0}^{N-1} [y(n) - s(n)]^2} \tag{3.7}$$

The root mean square error of the output signal is

$$\mathrm{RMSE} = \sqrt{\frac{\sum_{n=0}^{N-1} [y(n) - s(n)]^2}{N}} \tag{3.8}$$

Figure 3.9 shows the comparison of the time domain distribution of the signal before and after denoising. The signal-to-noise ratio and root mean square error of the signal before and after denoising are shown in Table 3.2.

According to the results, it can be seen that the EMD denoising algorithm based on the energy characteristics of the autocorrelation function is very effective, and the signal can be filtered on the premise of retaining the effective information of the original signal.

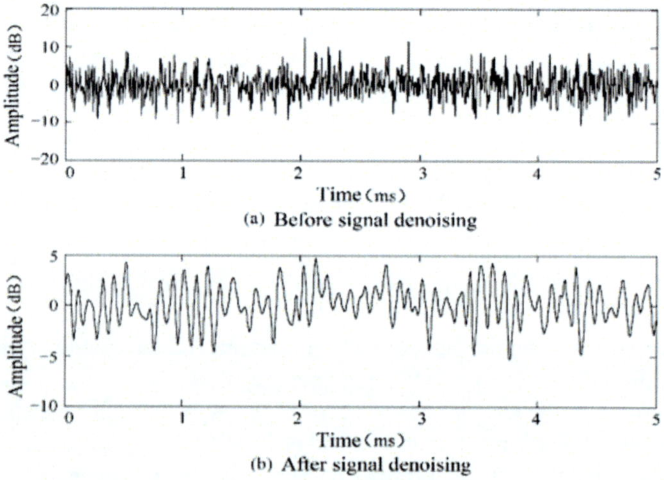

Fig. 3.9 Comparison of the time domain distribution of the signal before and after denoising

Table 3.2 Signal-to-noise ratio and root mean square error before and after denoising

	Signal-to-noise ratio (dB)	RMSE
Before signal denoising	2	0.71
After signal denoising	8.87	0.25

3.1.2 Self-Related Testing

Basic principles of self-related testing method

The autocorrelation detection method utilizes the uncorrelated characteristics between signal and noise, noise and noise, and performs autocorrelation operation on the input signal to improve the signal-to-noise ratio. The autocorrelation function $R_{xx}(m)$ of the signal $x(n)$ is

$$R_{xx}(m) = R_{ss}(m) + R_{SZ}(m) + R_{ZS}(m) + R_{zz}(m) \tag{3.9}$$

In the formula, $R_{ss}(m)$ and $R_{zz}(m)$ are the autocorrelation functions of the sinusoidal signal $s(n)$ and the noise signal $z(n)$, respectively, $R_{ZS}(m)$ and $R_{SZ}(m)$ are the cross-correlation functions of the noise signal $z(n)$ and the sinusoidal signal $s(n)$, respectively. The noise signal is randomly generated and has no correlation with the sinusoidal signal $s(n)$, so $R_{ZS}(m)$ and $R_{SZ}(m)$ are approximately 0. Gaussian white noise is non-zero only at the zero point and 0 at other points, which can be obtained

$$R_{zz}(m) = \sigma^2\delta(m) \tag{3.10}$$

In the formula, $\delta(m)$ is the unit sampling signal. The autocorrelation function $R_{ss}(m)$ of the sinusoidal signal is

$$\begin{aligned}
R_{ss}(m) &= \frac{1}{N-|m|}\sum_{n=0}^{N-1-|m|}[a\cos(\omega_0 n + \phi_0)][a\cos(\omega_0 n + \omega_0 m + \phi_0)] \\
&= \frac{a^2}{2}\cos(\omega_0 m) + \frac{1}{N-|m|}\left\{\frac{a^2}{2}\sum_{n=0}^{N-1-|m|}[a\cos(2\omega_0 n + \omega_0 m + 2\phi_0)]\right\}
\end{aligned} \tag{3.11}$$

When N approaches positive infinity, the second term in Eq. (3.11) approaches 0, the autocorrelation function $s(n)$ of the sinusoidal signal has the same frequency as the original signal, and it is also a sinusoidal signal. Therefore, $R_{xx}(m)$ can be simplified to

$$R_{xx}(m) \approx \frac{a^2}{2}\cos(\omega_0 m) + \sigma^2\delta(m) \tag{3.12}$$

It can be obtained from the above derivation that pre-processing the time domain signal (by performing autocorrelation operation on it) can reduce the influence of noise on the signal quality within a certain range. However, there is also an interference term in Eq. (3.12), so a constant will be generated when Fourier transform is performed on the time domain signal. In practical applications, discrete signals with limited length are processed, resulting in an unsatisfactory effect of an autocorrelation operation.

Assuming that the time domain waveform of the sinusoidal signal with noise is shown in Fig. 3.10, the signal of Gaussian white noise completely covers the sinusoidal signal, and the waveform after an autocorrelation operation of the signal is shown in Fig. 3.11. Comparing Fig. 3.10 and Fig. 3.11, it is found that the autocorrelation operation has a certain filtering effect on noise, but the effect is not ideal. Because the degree of time domain signal affected by noise is related to $|m|$, the larger $|m|$ is the greater the influence, so when the signal is relatively weak, the effect of an autocorrelation operation is not ideal.

The waveform of the signal after performing two autocorrelation operations is shown in Fig. 3.12. It can be seen from Fig. 3.12 that the valid signal is correctly extracted, and the signal-to-noise ratio of the signal can be improved by performing multiple autocorrelation operations.

Simulation and analysis

With the continuous development of science and technology, traditional data services in the network have lost their dominant position, and new services including voice and video gradually occupy a large amount of Internet bandwidth. A reasonable and

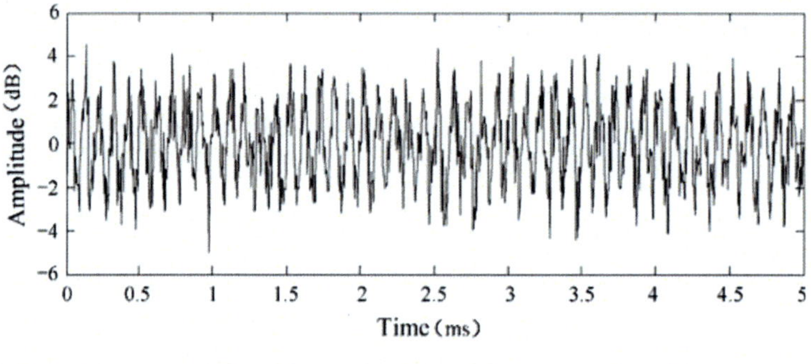

Fig. 3.10 Time-domain waveform of a noisy sinusoidal signal

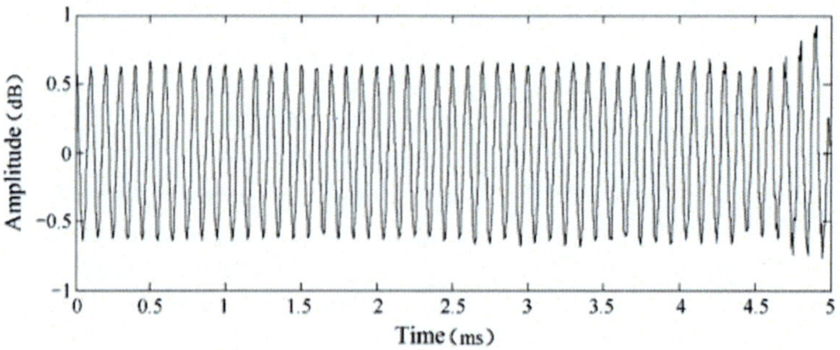

Fig. 3.11 The waveform of the signal after one autocorrelation operation

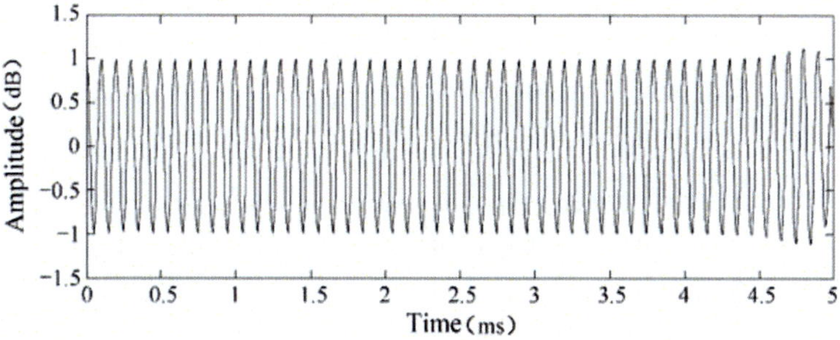

Fig. 3.12 The waveform of the signal after two autocorrelation operations

feasible business classification is an important guarantee to achieve traffic control and improve user satisfaction. Generally speaking, delay, jitter, and packet loss rate are three key parameters that affect user experience, and are often used as the main basis for dividing business categories to provide services according to actual needs.

At present, organizations such as ITU-T (International Telecommunication Union), ETSI (European Telecommunications Standardization Institute) and IETF (International Internet Engineering Task Force) have all conducted research on service classification and formulated a series of standards. This section mainly applies ITU-T standard. The signal-to-noise ratio is set to -5 dB, the sinusoidal signal is mixed with the noise signal, and the autocorrelation algorithm is used to simulate the denoising effect. Figure 3.13 shows the time domain signal and frequency domain signal before signal denoising, and Fig. 3.14 shows the time domain signal and frequency domain signal after signal denoising. As can be seen from Figs. 3.13 and 3.14, the sinusoidal signal is successfully screened out, the frequency of the effective signal before and after denoising is the same, the signal-to-noise ratio is still −5 dB, and the sinusoidal signal works frequency increased.

The autocorrelation detection method has a certain adaptive ability, and the denoising effect is not affected by the signal frequency. Compared with the EMD denoising algorithm, the autocorrelation detection method has a simple principle and a small amount of calculation, which is very suitable for the FSK system radar.

3.2 Speed Radar Angle Adaptive Algorithm

The radar speed measurement sensor adopts a non-contact method and uses the Doppler effect of the microwave signal relative to the ground speed to realize the speed measurement function. Because the propagation speed of electromagnetic waves in different environments is not very different and is less affected by weather, radar speed measurement has high reliability and stability. However, the train vibrates continuously during the traveling process, which causes the angle between the radar

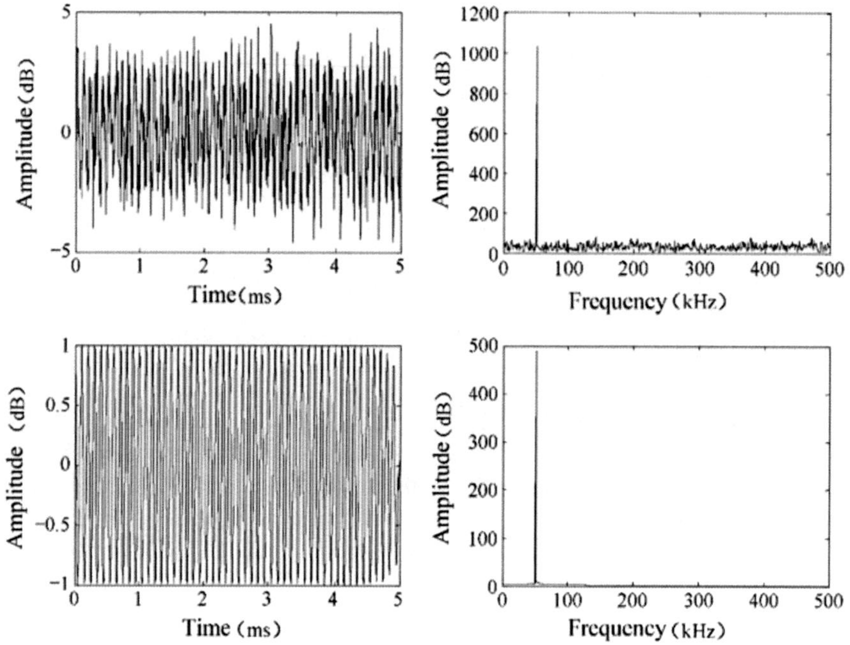

Fig. 3.13 The time domain signal and frequency domain signal before signal

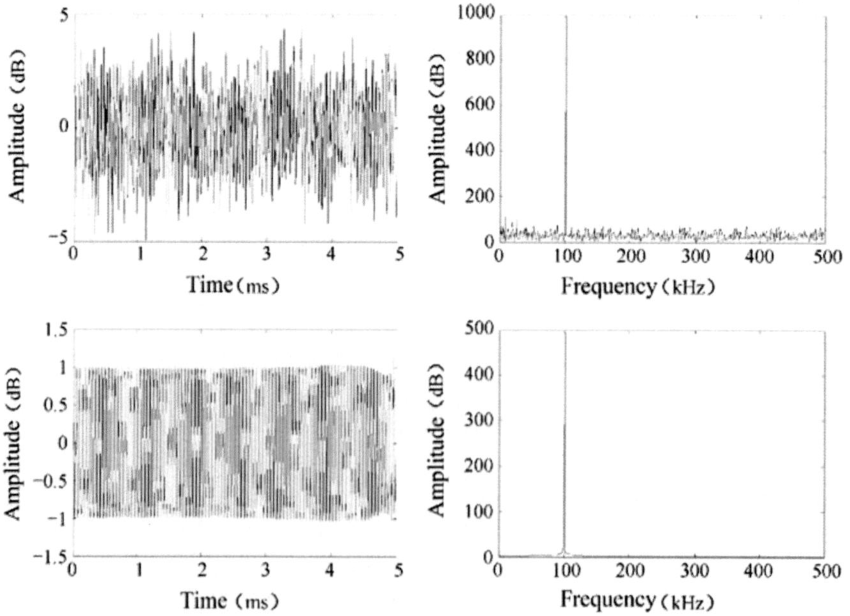

Fig. 3.14 The time domain signal and frequency domain signal after the signal is noisy

antenna and the ground to change, and the accuracy of the radar speed sensor decreases.

The traditional train speed measurement radar adopts the design idea of dual microwave modules, multi-AD sampling, and mutual conversion of various working modes under different circumstances to solve the problem of error and stability, or uses multiple radar sensors to co-locate and measure the speed. It has also been proposed to use a method based on adaptive Kalman filtering to improve the speed measurement accuracy of the radar.

The dual-antenna radar speed measurement sensor adopts a dual-antenna structure, and automatically corrects the angle between the antenna and the ground by using the angular relationship of the dual-antenna and the energy of the received signal. Through analyzing the data and testing with MATLAB, it is found that the speed measurement value obtained by the radar speed measurement sensor after the correction angle is closer to the real speed.

3.2.1 Train Speed Radar System

There are two ways to realize the train speed radar system, one is the same direction of the antenna, and the other is the different direction of the antenna. The structure of the train speed radar system is shown in Fig. 3.15.

The angle adaptive algorithm of the speed measuring radar adopts the structure of two antennas and four channels, and uses the Doppler frequency shift of the antenna 1 and the antenna 2 to obtain the actual angle between the antenna and the ground. The center frequency of radar operation is 24.15 GHz, and the radar speed measurement system transmits the processed speed information to the computer control system through the serial port. In practical applications, the angle self-correction algorithm can be used to eliminate the speed measurement deviation caused by the angle deviation. The algorithm is introduced and analyzed below.

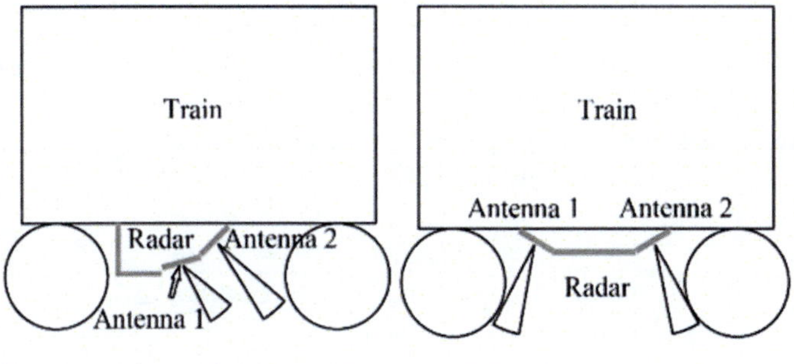

Fig. 3.15 The structure of the train speed radar system

Angle self-correction algorithm

The radar speed sensor utilizes the Doppler effect. The radar antenna installed on the train transmits electromagnetic waves with a fixed frequency to the ground and receives the signal reflected back from the ground. Because the train is in motion, there is a frequency difference between the transmitted signal and the echo signal, that is f_d, the Doppler frequency shift. The relationship between the speed of the train and the Doppler frequency shift is:

$$v = \frac{f_d c}{2 f_0 \cos\theta} \tag{3.13}$$

In the formula, v is the moving speed of the train, f_0 is the radar transmission frequency, and θ is the angle between the radar antenna and the ground.

When the antennas are in the same direction, the angle between the antenna 1 and the ground is θ_1, the angle between the antenna 2 and the ground is θ_2, and the angle difference between the antenna 1 and the antenna 2 is a fixed value. Let it be $\Delta\theta$, then the relationship between θ_1 and θ_2 is

$$\theta_2 - \theta_1 = \Delta\theta \tag{3.14}$$

The Doppler shifts of antenna 1 and antenna 2 are represented by f_{d1} and f_{d2} respectively, and the calculated velocities are v_1 and v_2 respectively, then there are

$$\begin{cases} v_1 = \dfrac{f_{d1} c}{2 f_0 \cos\theta_1} \\ v_2 = \dfrac{f_{d2} c}{2 f_0 \cos\theta_2} \end{cases} \tag{3.15}$$

Antenna 1 and Antenna 2 are mounted on a radar sensor, ideally with

$$v = v_1 = v_2 \tag{3.16}$$

The train vibrates during the movement, resulting in the deviation of the angle between the antenna and the ground, making $v \neq v_1 \neq v_2$, but the angle difference $\Delta\theta$ between the antenna 1 and the antenna 2 remains unchanged. Suppose the actual angle between antenna 1 and the ground is θ_1', then the actual angle between antenna 2 and the ground is $\theta_1' + \Delta\theta$, we can get

$$v = v_1' = v_2' = \frac{f_{d1} c}{2 f_0 \cos\theta_1'} = \frac{f_{d2} c}{2 f_0 \cos(\theta_1' + \Delta\theta)} \tag{3.17}$$

By formula (3.17), we can get

$$\frac{f_{d1}}{f_{d2}} = \frac{\cos\theta_1'}{\cos(\theta_1' + \Delta\theta)} \tag{3.18}$$

According to the formula of sum-difference product and $\sin(\theta_1')^2 + \cos(\theta_1')^2 = 1$, we can get

$$\cos\theta_1' = f_{d1}\sqrt{\frac{(\sin\Delta\theta)^2}{f_{d1}^2 + f_{d2}^2 - 2f_{d1}f_{d2}\cos\Delta\theta}} \tag{3.19}$$

Get the true speed of the train as

$$v = \frac{c}{2f_0}\sqrt{\frac{f_{d1}^2 + f_{d2}^2 - 2f_{d1}f_{d2}\cos\Delta\theta}{(\sin\Delta\theta)^2}} \tag{3.20}$$

When the antennas are in different directions, the calculation process is similar to that when the antennas are in the same direction. The difference is that when the antenna is anisotropic, the sum of the angles of antenna 1 and antenna 2 is a fixed value, and if it is θ_{sum}, then there is

$$\theta_1 + \theta_2 = \theta_{\text{sum}} \tag{3.21}$$

can be derived

$$v = \frac{c}{2f_0}\sqrt{\frac{f_{d1}^2 + f_{d2}^2 - 2f_{d1}f_{d2}\cos(\theta_{\text{sum}})}{(\sin\theta_{\text{sum}})^2}} \tag{3.22}$$

Test and simulation

The above algorithm is converted into a DSP program and programmed into the corresponding speed measurement prototype. The angle difference of the same direction antenna is 15°, and the angle sum of the anisotropic antenna is 90°. Install the radar on the train, and the real data collected by the host computer is shown in Fig. 3.16.

The radar installed on the train is connected to the PC through the serial line, and the data is collected by the host computer installed on the PC. The center frequency of the radar operation is 24.15 GHz, and the sampling frequency is 40 kHz. The number of time domain data points calculated by the DSP program per cycle is 1024 points, so its interruption time is: 1025/(40 × 103) = 25.6 ms. In order to facilitate the simulation, when collecting data, each group collects 400 cycles, a total of 1024 × 400 × 2 = 819,200 points (I road and Q road).

The radar transmits two signals whose frequencies are continuously converted. When the transmitted signal encounters the ground, it is reflected back to the radar and received by the antenna. Figure 3.17 shows the time domain signal and frequency

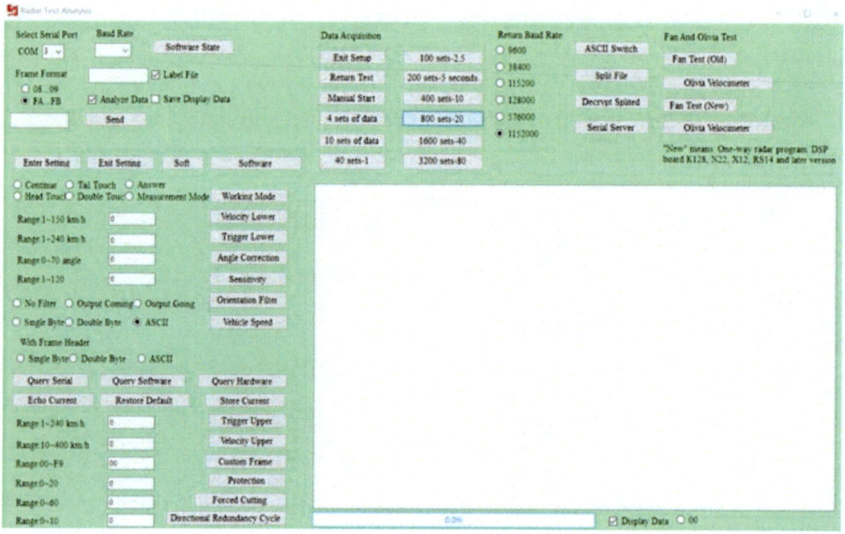

Fig. 3.16 Real data collected by the upper machine

domain signal corresponding to antenna 2 when the antennas are in the same direction. The phase difference between the I channel signal and the Q channel signal is 180°.

Collecting a large number of samples and analyzing them, it can be found that, without considering the angle deviation, the velocity calculated directly according to the Doppler principle has a large jump and is not stable enough; while the velocity obtained through self-correction is relatively stable and closer to the true velocity. Figure 3.18 shows the speed comparison of angle correction when the antennas are in the same direction and when the antennas are in different directions.

As can be seen from Fig. 3.18, the angle-corrected speed is closer to the true speed.

There are 4 prototypes of the antenna same-direction radar and 2 prototypes of the antenna anisotropic radar. The test image is shown in Fig. 3.19. In Fig. 3.19, the right display shows the speed of the antenna anisotropic radar, the left display shows the speed of the antenna same direction radar, and the lower part of the display is the real speed. The test found that the speed of the antenna anisotropic radar is closer to the real speed than that of the antenna same direction radar. Table 3.3 shows the speed deviation between the antenna co-directional radar and the antenna hetero-directional radar.

In train speed measurement, laser speed measurement has the highest accuracy, it works in the infrared and visible light bands, and the error is within 1 km/h. In this example, the speed measurement error of the antenna anisotropic radar is also within 1 km/h, but the laser speedometer can only work in a static state, and cannot be installed on a train and work in a vehicle-mounted dynamic mode, so it is impossible to measure the speed of the train in real time.

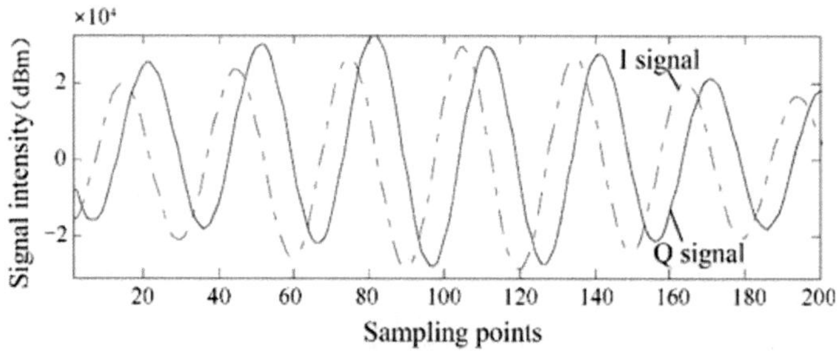

(a) The signal cycle is 1

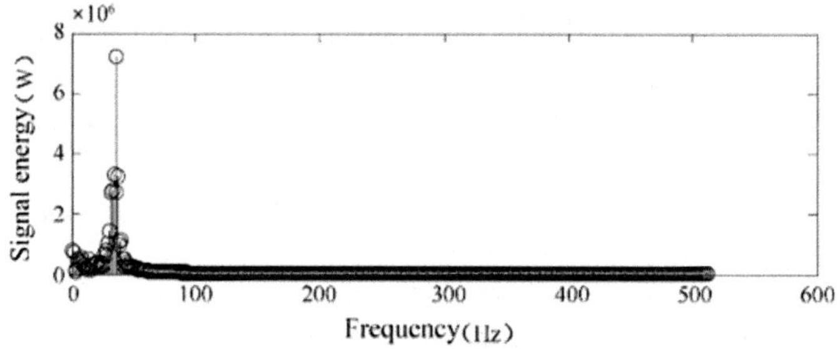

(b) Frequency domain signal after fast Fourier transformation

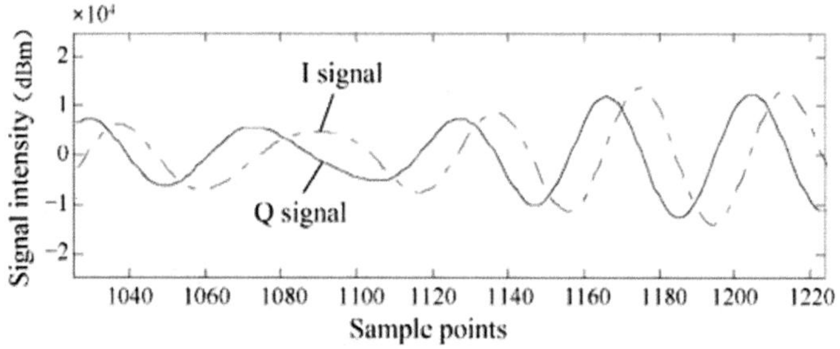

(C) The signal cycle is 2

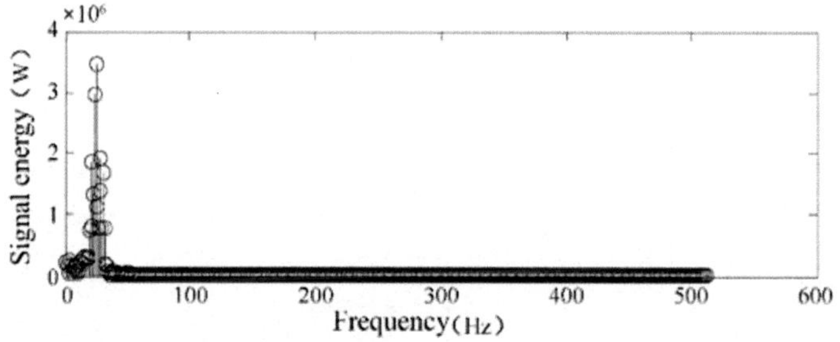

(d) Perform the frequency domain signal after the Fourier transform

Fig. 3.17 The antenna and the frequency domain signal corresponding to the time domain 2 in the same direction of the antenna 2

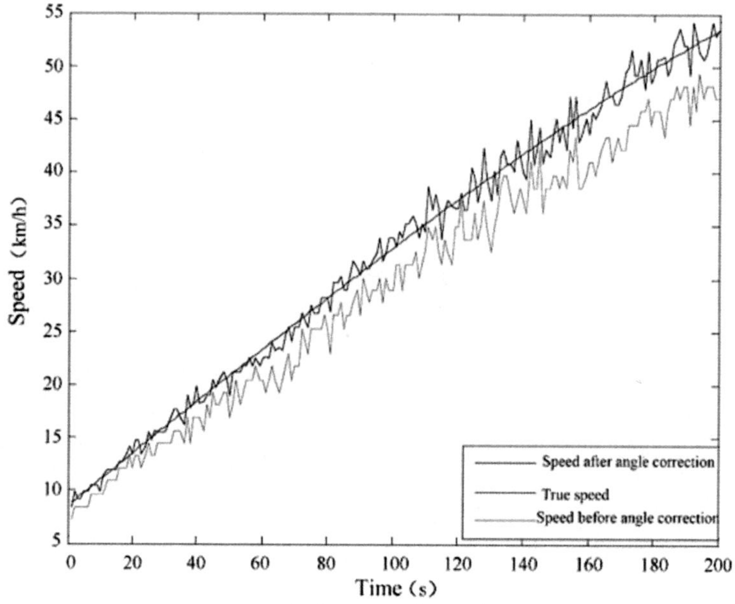

(a) Antenna anisotropic

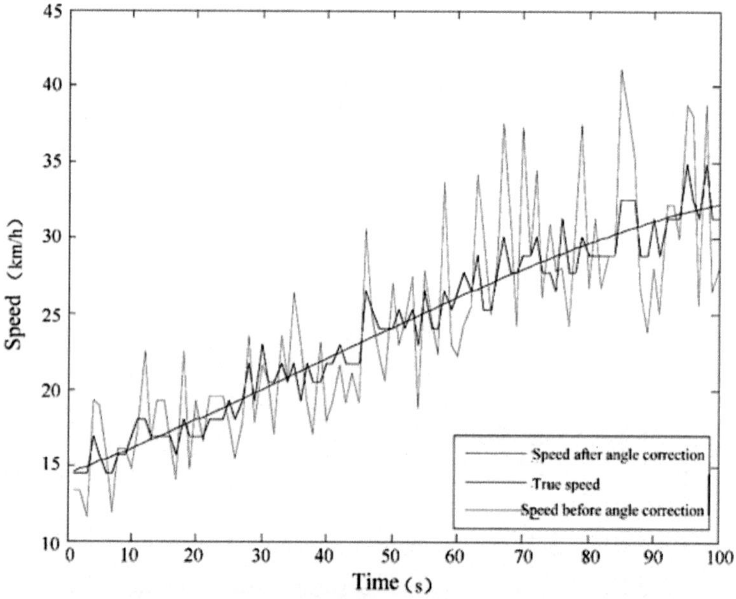

(b) Antenna homology

Fig. 3.18 Speed comparison of angle correction when the antennas are in the same direction and in different directions

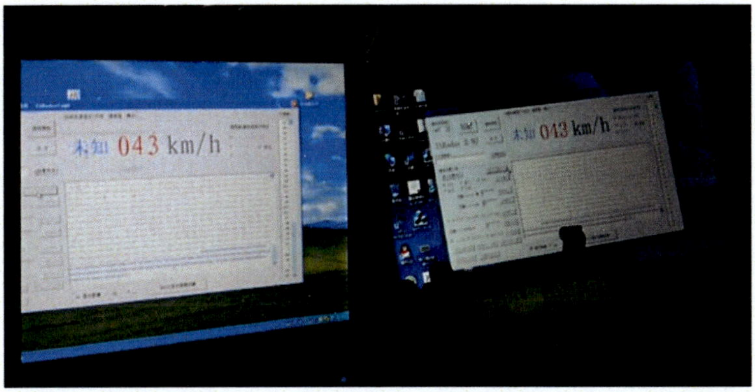

Fig. 3.19 Test image

Table 3.3 The speed deviation between the antenna and the heterogeneous radar of the radar and the antenna

	Antenna in the same direction				Antenna anisotropic	
	No. 1	No. 2	No. 3	No. 4	No. 1	No. 2
Deviation (km/h)	3~4	1~2	3~6	6~8	0~1	0~1
Fluctuation trend	High	Up and down	High	High	Up and down	Up and down

3.2.2 Amendment Algorithm Based on Radar Corners Based on Sample Statistical Characteristics

When using multi-target radar for traffic law enforcement, only by correctly judging the lane where the vehicle is located, can an effective snapshot image be obtained and used as the basis for traffic law enforcement. The traditional lane division method sets the number of lanes, lane width, radar installation position, radar swing angle, etc. in advance, and divides the lane by rotating the radar measurement coordinate system. The number of lanes, lane width, and radar installation position can all be measured accurately. Although the radar swing angle (the angle between the radar irradiation direction and the lane direction) can be obtained by measurement, various uncertain factors lead to low measurement accuracy and complicated operation. And this parameter has a great influence on the radar performance. If the radar swing angle is not suitable, there will be multiple shots, wrong shots, and missed shots, which will seriously affect the radar performance.

This section designs a radar swing angle self-correction algorithm, so that the radar can calculate the radar swing angle by itself through the collected target signal during use, and simplify the algorithm complexity of later signal processing.

Characteristics

Since the actual traffic situation is more complicated, it can be assumed that the lanes monitored by the radar are parallel and most of the vehicles travel along the lanes. If there is a certain swing angle when the radar is installed, the trajectory of each vehicle passing through the radar irradiation area is a straight line inclined at a certain angle relative to the radar. If the inclination angle of this line is known, the radar swing angle is obtained.

Statistics on the motion trajectories of a large number of vehicles in a certain period of time are used to obtain the distribution of vehicle trajectories and analyze their characteristics. Figure 3.20 shows the strong signal point cloud in the radar irradiation area for a certain period of time.

In Fig. 3.20a, three trajectories that are approximately straight lines can be clearly seen, which correspond to the three lanes actually monitored by the radar. In addition, there are ghost images composed of clutter and passing signals in the non-monitored area. Most clutter signals can be filtered out by setting the amplitude threshold.

Because the vehicle travels in the direction of the lane, the direction of the motion trajectory of the vehicle represents the direction of the lane. If the direction of the radar illumination is parallel to the direction of the lane, then the trajectory of the vehicle in the trajectory graph should be vertical. But in fact, in most cases, due to the non-standard installation, there is always a certain angle between the direction of the radar irradiation and the direction of the lane, resulting in the trajectory of the vehicle is not vertical. At this time, the angle between the vehicle's trajectory and the vertical direction is the radar swing angle.

The judgment of the vertical state of the vehicle trajectory is a manual process. Therefore, in engineering implementation, it is necessary to find an algorithm that can automatically judge the vertical state, and obtain the rotation angle by comparing with the original state.

Feature extraction and algorithm implementation

Assuming that most of the vehicles are driving along the lane, it can be considered that there are enough vehicles passing by in a long enough time, and there are the most points on the lane, so how to get the direction when the points are the most?

As can be seen from Fig. 3.20, all points are distributed on the 2D plane where the road surface is located. In order to facilitate statistics, the point set of the two-dimensional plane is converted to the one-dimensional coordinate axis, that is the point (x_i, y_i), $i \in R$ is projected onto the axis to obtain the point x_i, $i \in R$. At the same time, rotate the trajectory at different angles, count the number of points that fall into the axis unit interval after each rotation, and obtain the distribution of points $\{a_1, a_2, \cdots, a_n\}$, $n \in N$, and the specific distribution can be displayed by a histogram.

The distribution of projection points is shown in Fig. 3.21. Figure 3.21a is the distribution of projection points in the vertical state. The distribution of projection points is very dense. Figure 3.21b is in the non-vertical state. The distribution of projected points, the distribution of projected points is relatively scattered. It can

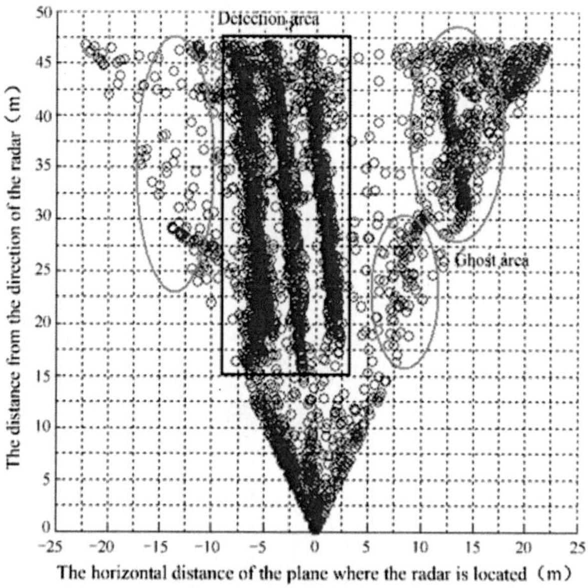

(a) Track map of the original signal

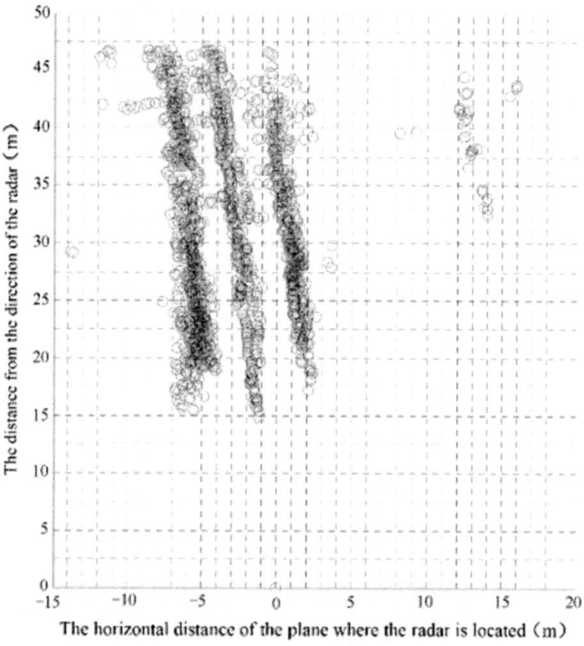

(b) Track plot of the preprocessed signal

Fig. 3.20 A strong signal point cloud map in the radar irradiation area during a certain period be seen that the trajectory when the projection points are the densest rotates to the vertical state.

The number of points in a certain time period is a fixed value, i.e.

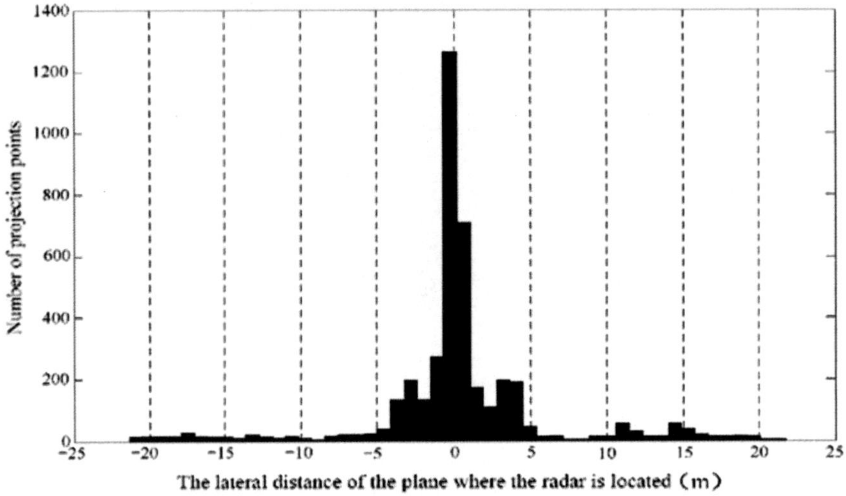

(a) The distribution of projection points in vertical state

(b) The distribution of projection points in non-vertical state

Fig. 3.21 Projection point distribution situation. The distribution of projection points (continued)

$$a_1 + a_2 + \cdots + a_n = \mathrm{C} \tag{3.23}$$

In the formula, is a constant. To get the aggregation degree of projected points under different rotation angles, it is necessary to under the premise of formula (3.23), solve the maximum value of the following objective function

$$f(a_1, a_2, \cdots, a_n) = a_1^2 + a_2^2 + \cdots + a_n^2 \tag{3.24}$$

Obviously, this is an extreme value problem with constraints, which can be solved by the method of Lagrange multipliers.

Build Lagrangian Functions

$$L(a_1, a_2, \ldots, a_n, \lambda) = f(a_1, a_2, \ldots, a_n) + \lambda\varphi(a_1, a_2, \ldots, a_n) \tag{3.25}$$

In the formula, λ is the Lagrange multiplier, and the constraint condition is

$$\varphi(a_1, a_2, \ldots, a_n) = a_1 + a_2 + \cdots + a_n - \mathrm{C} = 0 \tag{3.26}$$

The system of equations consisting of the partial derivatives of the Lagrangian function can be obtained

$$\begin{cases} L_{a_1} = f_{a_1} + \lambda\varphi_{a_1} = 0 \\ L_{a_2} = f_{a_2} + \lambda\varphi_{a_2} = 0 \\ \vdots \\ L_{a_n} = f_{a_n} + \lambda\varphi_{a_n} = 0 \\ L_\lambda = \varphi(a_1, a_2, \ldots, a_n) = 0 \end{cases} \tag{3.27}$$

Solving Eq. (3.27), the stable maximum point can be obtained, and the lengthy calculation process is omitted here.

The algorithm implementation process is as follows.

(1) Use multi-target radar to collect traffic data within a certain period of time to obtain a trajectory map.
(2) Rotate the orbit at different angles, project it on the x-axis, and count the number of points that fall into the unit interval of the x-axis after each rotation.
(3) Calculate the sum of squares of the points that fall into these intervals.
(4) Compare the sum of squares of points at different rotation angles. The state corresponding to the maximum value of the sum of squares is the vertical state of the vehicle trajectory, and the corresponding rotation angle is the radar swing angle.

Coordinate conversion

The initial target information is obtained in the polar coordinate system, but the polar coordinate system is not suitable for algorithms such as filtering and tracking in the

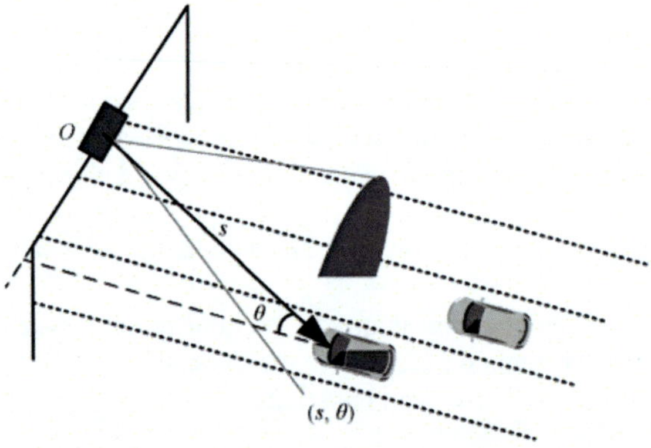

Fig. 3.22 Polar coordinate system where the target is located

later data processing of the radar. Therefore, it is necessary to update the speed, distance, angle and other information of the radar through coordinate conversion technology, and put them into the same coordinate system for processing.

The polar coordinate system where the target is located is shown in Fig. 3.22. For a target, the following information can be given:

v is the radial velocity of the target relative to the radar movement, θ is the radial distance between the target and the radar, and is the included angle between the target and the radar irradiation direction.

In fact, in most cases, the vehicle moves in a straight line along the lane. If the target information needs to be judged by lane, it is most effective to use the Cartesian coordinate system.

Taking the position of the radar as the origin of the coordinates, and taking the projection of the radar irradiation direction (the direction of the beam center) as the y-axis, a Cartesian coordinate system is established. The Cartesian coordinate system where the target is located is shown in Fig. 3.23.The conversion relationship between Cartesian coordinates and polar coordinates is:

$$\begin{cases} x = s \sin\theta \\ y = s \cos\theta \end{cases} \tag{3.28}$$

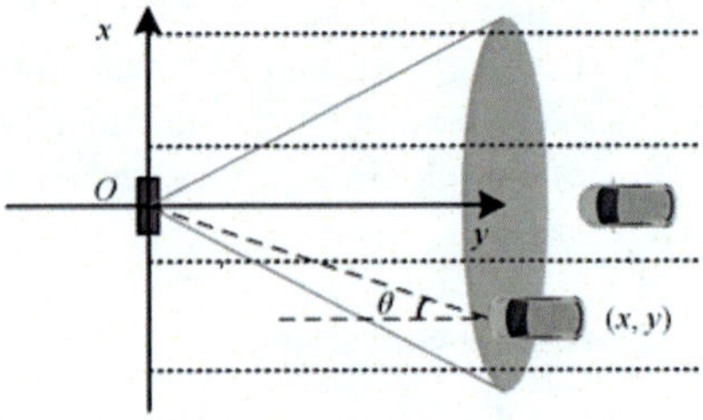

Fig. 3.23 Cartesian coordinate system where the target is located

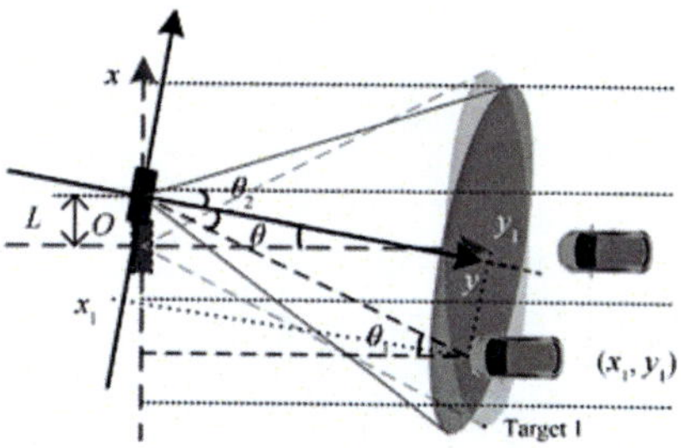

Fig. 3.24 Offset Cartesian coordinate system

In practical applications, there is often an included angle between the projection of the radar illumination direction and the direction of the lane, and the radar may also deviate from the center of the lane, so a more general situation needs to be considered. The offset Cartesian coordinate system is shown in Fig. 3.24.

For ease of understanding, the ideal Cartesian coordinate system with the lane direction as the y-axis is called the lane Cartesian coordinate system, and the actual coordinate system is called the radar Cartesian coordinate system.

Assuming that the actual installation position of the radar has a distance L offset from the origin O of the lane Cartesian coordinate system, the angle between the radar irradiation direction and the lane direction is, that is, the radar swing angle.

At this time, through the calculation of the radar phase difference, the angle θ between the target 1 and the y-axis of the radar rectangular coordinate system can be directly obtained. For the lane Cartesian coordinate system, $\theta_1 = \theta + \theta_2$. Assuming that the radar swing angle can be obtained by measurement, the distance between target 1 and the radar is s, then

$$\begin{cases} x = s\sin\theta_1 - L = s\sin(\theta + \theta_2) - L \\ y = s\cos\theta_1 = s\cos(\theta + \theta_2) \end{cases} \tag{3.29}$$

In the radar Cartesian coordinate system, the coordinate of target 1 is (x_1, y_1), then the coordinate conversion relationship between in the lane Cartesian coordinate system (x, y) and in the radar Cartesian coordinate system (x_1, y_1) is:

$$\begin{cases} x = x_1\sin\theta_2 - y_1\sin\theta_2 - L \\ y = y_1\sin\theta_2 - x_1\sin\theta_2 \end{cases} \tag{3.30}$$

Since the later processing of radar data is carried out in the lane rectangular coordinate system, after the coordinate transformation of the information (v, s, θ) measured by the actually installed radar, the obtained information (v', s', θ') is:

$$\begin{cases} v' = v\cos(\theta + \theta_2) \\ s' = s\cos(\theta + \theta_2) \\ \tan\theta' = \frac{s\cos(\theta+\theta_2)}{s\sin(\theta+\theta_2)-L} \end{cases} \tag{3.31}$$

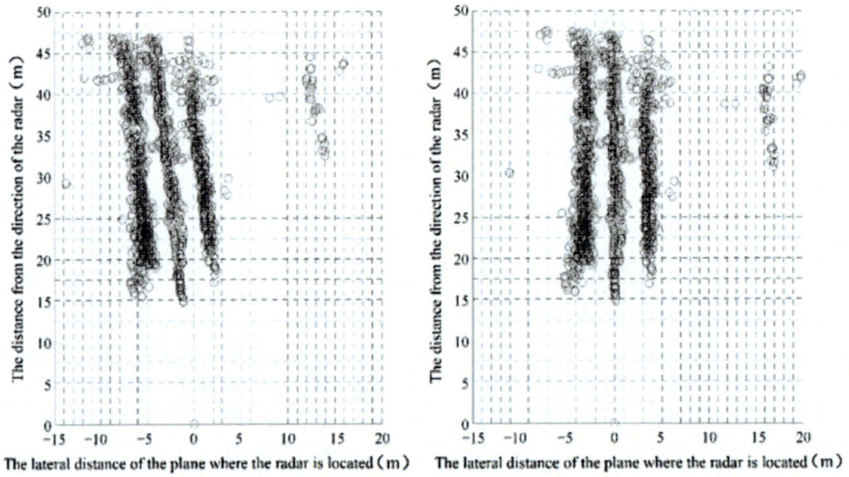

Fig. 3.25 Correction effect comparison

Simulation and analysis

The above algorithm is simulated, and the comparison of trajectory correction effect is obtained as shown in Fig. 3.25.

It can be seen that the trajectory of the vehicle with a certain inclination angle is corrected to become a vertical state. Figure 3.26 shows the change of the statistical sum of squares. The maximum value appears at the densest projection point, that is, the vertical state of the vehicle trajectory. At this time, the rotation angle is 5.25°, which is the radar swing angle.

It can be found that the calculated results are basically consistent with the actual results, obtained through the actual road test, and a good capture effect can be obtained by using the swing angle to capture.

3.3 Summary

This chapter introduces two denoising algorithms, both of which are independent of the signal frequency and have adaptive capabilities. The autocorrelation detection method has a small amount of calculation, a simple working principle and a good denoising effect, and can obtain a signal with the same frequency as the sinusoidal signal, so it is very suitable for FSK system radar (only the frequency information of the effective signal needs to be obtained).The empirical mode decomposition method based on the autocorrelation function, using the energy characteristics of the autocorrelation function of the noise and the effective signal, can accurately distinguish

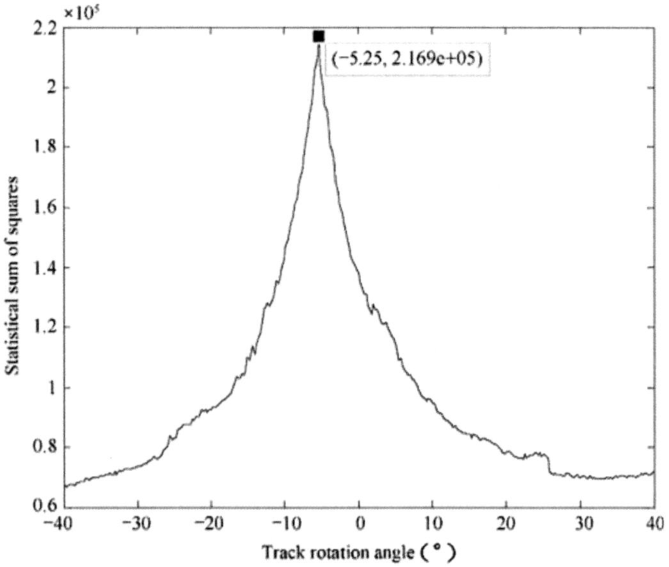

Fig. 3.26 Statistics square and change diagram

the IMF component dominated by the useful signal and the IMF component dominated by the noise, and recombine the IMF components of the useful signal, it can effectively filter out the noise in the original signal.

This chapter also introduces the same-direction and anisotropic antenna structures and speed-measuring principles of the train speed-measuring radar, and compares the speed-measuring performance of the two structures. The dual-antenna train radar speed measurement algorithm only adds a low-cost antenna to the single-antenna Doppler speed measurement radar. By using the relative angle between the dual antenna and the ground for detection and calculation, it not only corrects the speed measurement error caused by the angle deviation, It also guarantees the stability and reliability of the radar speed measurement system.

Aiming at the problem of the swing angle error of traffic speed radar in practical applications, this chapter carries out feature extraction on the basis of a large number of data feature analysis, and proposes a radar swing angle self-correction algorithm based on the statistical characteristics of samples, optimized the problem of low capture rate caused by swing angle error.

Chapter 4
Target Recognition and Tracking of Microwave Velocity Radar

In the previous introduction of traffic radar, it is pointed out that the current traffic radar can only measure speed, and the effective capture rate is low, and the idea of the CW-FMCW composite system is given. In this chapter, aiming at the problems of velocity attenuation and distance fluctuation caused by system noise and measurement methods in practical application, the Kalman filtering is used to denoise the target dynamic information, compensate for the speed and correct the distance. In addition, because the radar algorithm is finally implemented in the digital signal processing module, regarding the idea of the CW-FMCW composite system, the original single system speed measurement algorithm is replanned, and the algorithms of the two systems in the CW-FMCW composite system are time-sharing multiplexed, and the selection rules in the speed fusion process are briefly described. Through the complete radar system test, the feasibility and effectiveness of the improved algorithm are verified.

4.1 Algorithm Optimization of Single Target Radar

4.1.1 Optimization of Velocity and Distance Measurement Algorithm Based on Kalman Filter

For radar, the distance and velocity information of the target are the most critical, but the distance and velocity information measured by radar contain noise. In practical application, the relative position of the vehicle and radar is shown in Fig. 4.1.

In Fig. 4.1, s_m is the distance between vehicle and radar measured by radar, V_m is the speed of the vehicle measured by radar, V is the actual speed of the vehicle, h is the vertical distance between radar and vehicle. The relationship between V_m and V is

L. Cao et al., *Target Recognition and Tracking for Millimeter Wave Radar in Intelligent Transportation*, https://doi.org/10.1007/978-981-99-1533-0_4

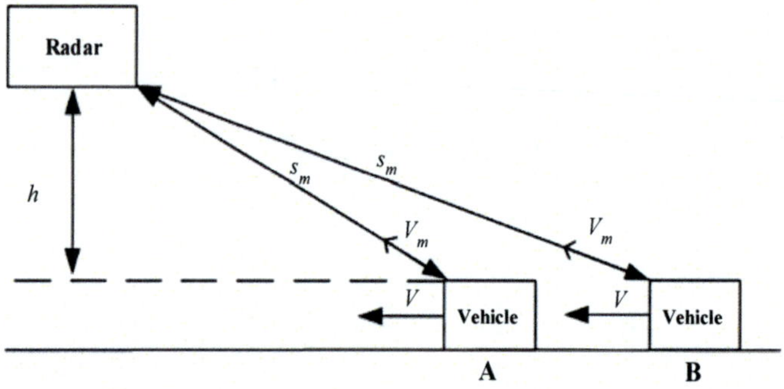

Fig. 4.1 Relative position of the vehicle to radar

$$V_m = V\sqrt{1 - \frac{h^2}{s_m^2}} \tag{4.1}$$

The radar interruption period t_0 is 13.12 ms, and the time of vehicle passing through the radar irradiation range is very short. Therefore, it can be considered that V remains unchanged during the speed measurement. From the formula (4.1), V_m gradually decreases.

Considering the velocity state value of the previous time obtained by radar and the velocity measurement value of the current time, the noise is filtered by combining the time correlation of velocity and distance, and the velocity is compensated and the distance is corrected to obtain more accurate distance and velocity information.

Building a mathematical model and assuming that the velocity is v and the position of the target at time k is $s(k)$, then after sampling time t, the position of the target is

$$s(k+1) = s(k) + vt \tag{4.2}$$

Initialize parameters, get state equation and observation equation of target

$$\begin{cases} \boldsymbol{X}(k+1) = \boldsymbol{A}\boldsymbol{X}(k) + \boldsymbol{W}(k) \\ \boldsymbol{Y}(k) = \boldsymbol{H}\boldsymbol{X}(k) + \boldsymbol{U}(k) \end{cases} \tag{4.3}$$

In the formula, $\boldsymbol{X}(k) = [s(k)v(k)]^{\mathrm{T}}$ represents the initial values of speed and distance; $\boldsymbol{A}$ is a state transition matrix; $\boldsymbol{H}$ is the observation matrix; $\boldsymbol{P}(k)$ is the covariance matrix of $\boldsymbol{X}(k)$; $\boldsymbol{W}(k)$ and $\boldsymbol{U}(k)$ are input white noise and observation noise, and their covariance matrices are $\boldsymbol{Q}$ and $\boldsymbol{R}$, respectively.

Since V_m gradually decreases, the classical Kalman filtering algorithm is improved below. Define the compensation coefficient $\eta(k)$, the initial value $\eta_0 = 1$, and the updated equation of the compensation coefficient (4.4). It can be seen that the state transition matrix $\boldsymbol{A}$ and the observation matrix $\boldsymbol{H}$ are changed, including the compensation coefficient $\eta(k)$, which is equivalent to the speed and distance correlation and filtering.

$$\eta(k+1) = \frac{1}{\sqrt{1 - \frac{h^2}{s(k+1)^2}}} \approx 1 + 0.5\left[\frac{h}{s(k+1)}\right]^2 \tag{4.4}$$

It can be seen from Formulas (4.4) that the compensation coefficient is related to the state $s(k)$ of the filter. The corresponding state transition matrix and observation matrix can be expressed as

$$\begin{cases} \boldsymbol{A} = \begin{bmatrix} 1 & -\frac{t_0}{\eta(k)} \\ 0 & 1 \end{bmatrix} \\ \boldsymbol{H} = \begin{bmatrix} 1 & 0 \\ 0 & \frac{1}{\eta(k)} \end{bmatrix} \end{cases} \tag{4.5}$$

Different Kalman filter iterative initial values are set according to the actual experience. By comparing the filtering results, it is found that when the iterative initial value is shown in the form (4.6), the filtering compensation effect is the best.

$$\begin{cases} \boldsymbol{P}(k) = \begin{bmatrix} 0.5^2 & 0 \\ 0 & 0.3^2 \end{bmatrix} \\ \boldsymbol{Q} = \begin{bmatrix} 0.02^2 & 0 \\ 0 & 0.01^2 \end{bmatrix} \\ \boldsymbol{R} = \begin{bmatrix} 0.1^2 & 0 \\ 0 & 0.22^2 \end{bmatrix} \end{cases} \tag{4.6}$$

In this algorithm, the physical meanings of $s(k)$ and v are the same as those of s_m and V, respectively. The output result v of the filter is the vehicle speed V. The Kalman filter continuously predicts and corrects the variance, and recursively updates the compensation coefficient, to obtain more accurate distance and speed information.

4.1.2 DSP Algorithm Improvement Based on CW-FMCW Hybrid System

The center frequency of high precision radar for test is 24.15 GHz. The digital signal processing uses TI's TMS320C6713B processor, which can execute eight 32-bit instructions per cycle, and can effectively meet the real-time requirements of radar signal processing.

To make full use of the advantages of the composite system and adjust the hardware circuit, the DSP algorithm should be considered. According to the information contained in the triangular wave component and the DC component in the CW-FMCW composite system, the reprogramming algorithm: due to the high demand for speed measurement accuracy in the traffic field, the speed measurement is realized by the CW system; because of the need for high-precision target positioning and tracking, so the distance measurement using FMCW system; the velocity under CW system should be used to replace the velocity under FMCW system according to certain rules, and the velocity and distance information should be compensated and corrected by Kalman filtering.

According to the above analysis, based on the original single-system velocity measurement algorithm, the algorithms of the two systems in the CW-FMCW composite system are time-sharing multiplexing. The re-planned composite system algorithm flow is shown in Fig. 4.2.

In addition to the parallel processing of the signals of the two systems, the re-planned composite constitution algorithm also applies the speed comparison function to make choices in the process of speed fusion, and the rules are as follows.

(1) If the triangular wave has no velocity, the CW signal should be cleared even if it has velocity. In this way, the current situation that the target cannot be distinguished in the case of short rear vehicle distance and close speed can be well avoided and tracked as a target.
(2) If the triangular wave has velocity and the CW signal has no velocity, the radar will output the distance value, but the speed should be zero. Because the radar may irradiate the stationary target, leading to a triangular wave to determine the false target.
(3) If the velocity of the triangular wave is greater than 50 km/h, the difference between the CW signal and its velocity is more than 30 km/h, there must be one of the two waves, which will be judged as abnormal and marked.
(4) The speed of triangular wave correction in other cases.

After the velocity is obtained according to the above rules, the velocity is compensated by the Kalman filter, and the distance fluctuation is corrected so that the velocity measurement and distance measurement do not interfere with each other.

4.1.3 Simulation and Analysis

The feasibility of the algorithm is verified by road test data acquisition. The interrupt period of the test radar is about 13 ms, with 1024 points collected per interrupt period. Time domain and frequency domain signals are shown in Fig. 4.3.

1. Velocity and distance comparison before and after Kalman filtering

Kalman filter speed compensation effect as shown in Fig. 4.4, Kalman filter distance correction effect as shown in Fig. 4.5.

Fig. 4.2 The re-planning algorithm flow of the composite system

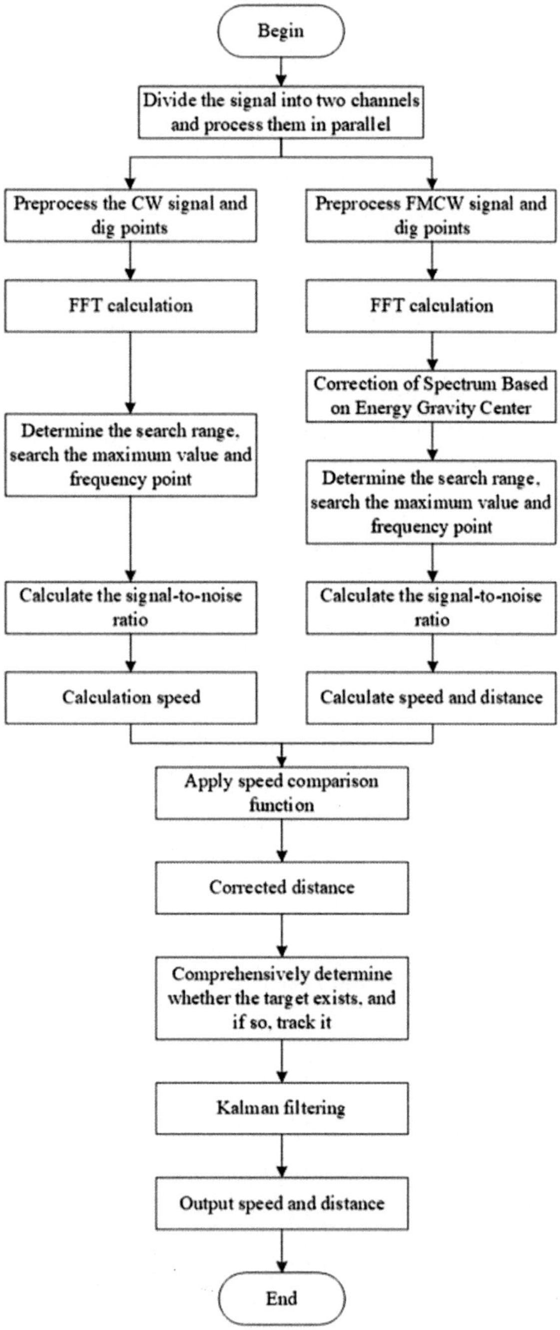

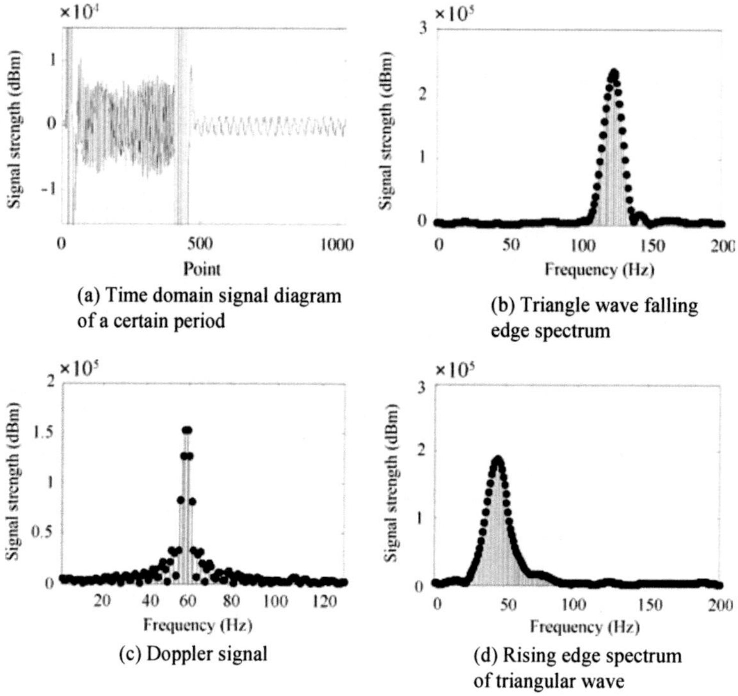

(a) Time domain signal diagram of a certain period
(b) Triangle wave falling edge spectrum
(c) Doppler signal
(d) Rising edge spectrum of triangular wave

Fig. 4.3 Time domain signal and frequency domain signal

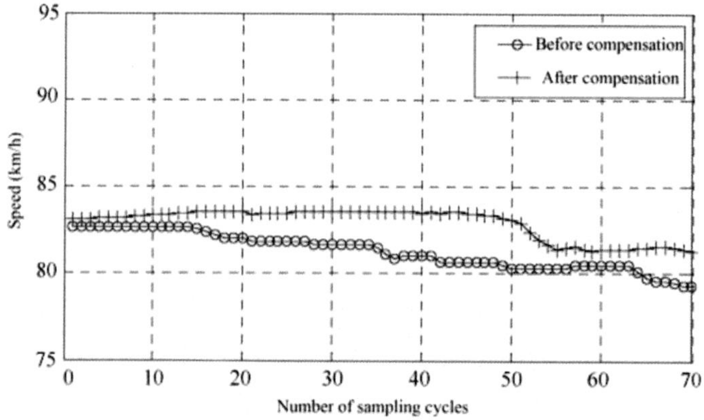

Fig. 4.4 Kalman filter speed compensation effect

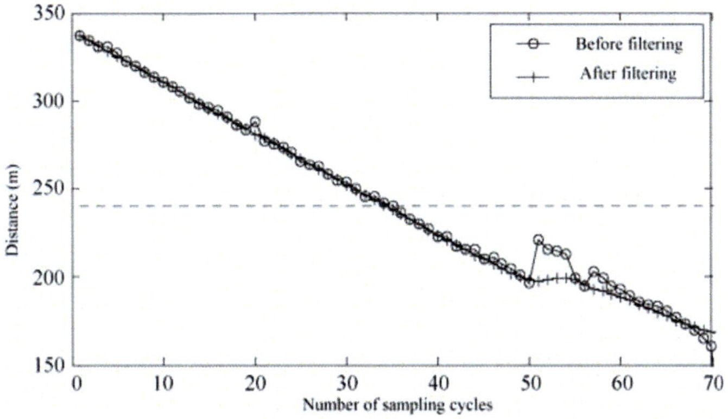

Fig. 4.5 Kalman filter distance correction effect

Figure 4.4 shows that the velocity fluctuation range after compensation is 0–2.25 km/h, which is less than that before compensation (0–3.41 km/h). The compensated target maintains uniform motion, but the speed drops suddenly after the 50th cycle, and the reason can be found in Fig. 4.5. The larger distance fluctuation near the cycle leads to the larger recursive variance, which affects the speed compensation.

Kalman filter can compensate for the speed well and filter out the random noise of the distance to some extent, so that the measurement and tracking results of the target are optimal.

2. Comparison of velocity measured by the composite system and single system

Field road test data simulation results are shown in Fig. 4.6.

Figure 4.6a, c denote the variation of distance with time, and an envelope represents a vehicle, which is measured by the FMCW system and corrected by the composite system. Figure 4.6b, d denote the speed of the same vehicle measured under three systems. It can be seen that in the FMCW system, the measured velocity attenuation is serious and the error is large; in the CW system and composite system, the measured velocity is consistent, but there is also a small attenuation.

It can be seen from Fig. 4.6d that the speed curves of the two vehicles are connected unthe der CW system, and the vehicle situation cannot be distinguished. The FMCW system and the composite system determine two vehicles. This is because the CW system is sensitive to moving targets and has a long time to follow the vehicle, so it is difficult to distinguish targets when the vehicle is dense and the speed is close.

Overall, the velocity measured in the composite system is the most stable, because the Kalman filter has the effect of compensation and correction, the results are closer to the real trajectory of the target.

3. Effective capture rate

The effective capture rate of the composite physique radar is shown in Table 4.1 by comparing the captured video with the captured image.

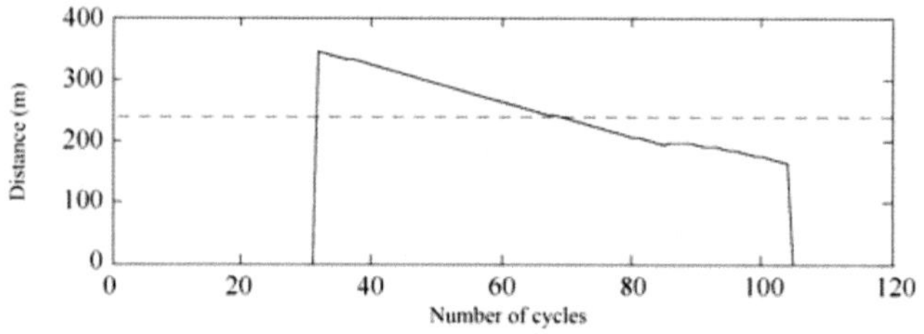

(a) Simulation result 1 - Distance

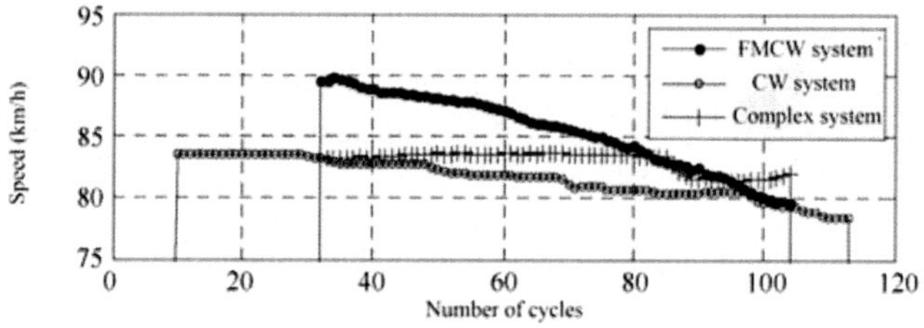

(b) Simulation result 1 - Speed

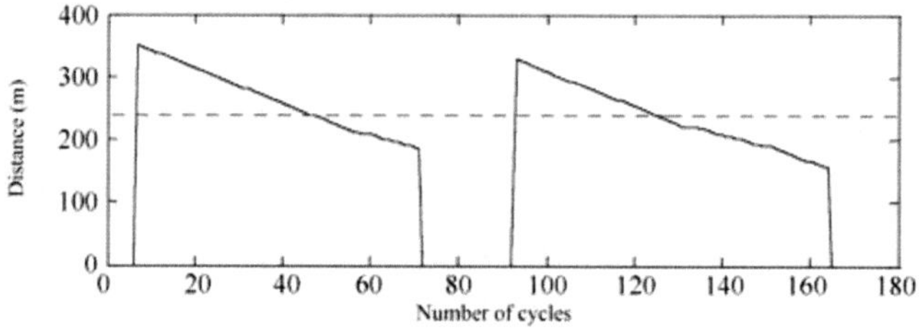

(c) Simulation result 2 - Distance

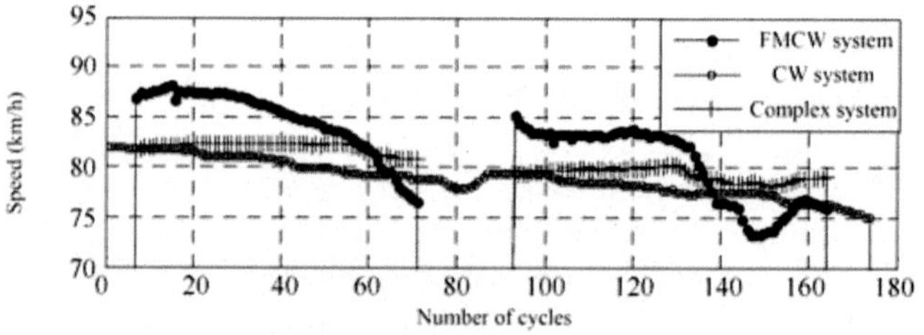

(d) Simulation result 2 - Speed

Fig. 4.6 Simulation results of on-site drive test data

Table 4.1 Effective capture rate of composite radar

	Group 1	Group 2	Group 3	Group 4
The actual number of vehicles passing	258	318	334	289
Number of shots	261	320	334	295
Missing shots of a car	1	0	0	2
The continuous car missed shots	1	3	2	0
Multiple shots	5	5	2	8
Number of valid shots	256	315	332	287
Effective capture rate (%)	99.22	99.06	99.40	99.31

As can be seen from Table 4.1, the effective capture rate of the composite system radar exceeds 99%, and the effective capture rate of the CW system radar commonly used in the market can only reach 90%. Therefore, the composite system radar has broad application prospects.

4. Capture effect

The CW system radar cannot measure the distance information of the vehicle, and only judges whether to trigger the capture according to whether the intensity of the target echo signal reaches the threshold value. This method leads to the phenomenon of car rushing and long-distance undershooting. Figure 4.7 shows the capture effect of the CW system radar. Figure 4.7a is a normal distance snapshot; Fig. 4.7b is a rushing car, and the valid information of the license plate cannot be obtained; Fig. 4.7c shows the situation where the license plate is not captured under long-distance conditions, which lead to waste.

The composite system radar has accurate ranging, and can be set to trigger a snapshot at a fixed distance, and the capture effect is good. The capture effect of the composite system radar is shown in Fig. 4.8.

It can be seen from Fig. 4.8 that the capture is triggered at the same position, and effective information such as license plate, vehicle type, and driver can be obtained. The capture effect is good, and it is not affected by vehicle type, position, and speed.

(a) (b) (c)

Fig. 4.7 Snapshot effect of CW system radar

Fig. 4.8 Snapshot effect of composite radar

4.2 FSK Radar Speed Measurement Algorithm

4.2.1 FSK Radar Speed Measurement Principle

The FSK radar is upgraded based on the traditional Doppler radar, and the basic working principle is still Doppler frequency shift. However, the transmitted signal of the radar is not a continuous wave with a fixed frequency, but a square wave with frequency modulation. The electromagnetic wave emitted by the radar has two frequencies, and the signal has a time difference. The electromagnetic wave is reflected when it meets the target and receives ved by the antenna. The radar uses the phase difference between the two signals with different frequencies to determine the distance of the target.

1. Single target recognition for FSK radar

The relationship between the modulated square wave frequency and time is shown in Fig. 4.9.

The voltage waveform the of FSK radar modulation signal is a square wave with the same period and duty cycle. When measuring the speed, the radar emits a square wave (the frequency at low level V_1 is f_1, and the frequency at high level V_2 is f_2). The transmitted signal can be expressed as

$$s(t) = \begin{cases} A\cos(2\pi f_1 t + \phi) \\ A\cos(2\pi f_2 t + \phi) \end{cases} \tag{4.7}$$

When there is a target in front of the radar, the echo signal is

$$s_r(t) = Ks(t - t_r) = KA\cos[\omega_0(t - t_r) + \phi] \tag{4.8}$$

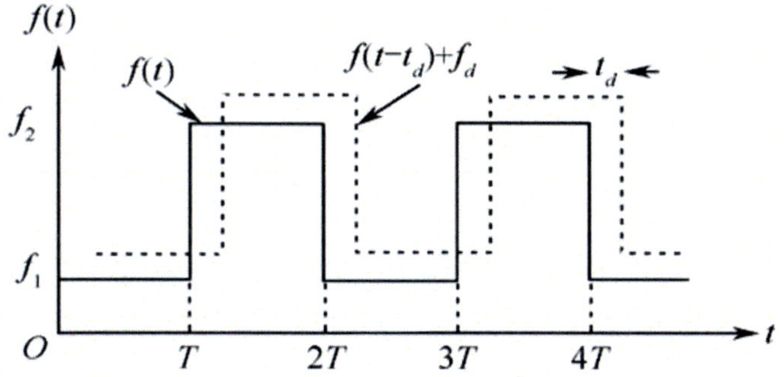

Fig. 4.9 Relationship between frequency and time of modulating square wave

$$s(t) = \begin{cases} KA\cos[2\pi f_1(t - t_r) + \phi] \\ KA\cos[2\pi f_2(t - t_r) + \phi] \end{cases} \tag{4.9}$$

Signal delay time t_r is very short and negligible. It can be seen from Formulas (4.6) that the Doppler frequency shift corresponding to two signals with different frequencies is

$$f_{d1} = \frac{1}{2\pi}\frac{\mathrm{d}\phi_1}{\mathrm{d}t} = \frac{1}{2\pi} \times (2\pi\frac{2}{\lambda_1}\upsilon_r) = \frac{2}{\lambda_1}\upsilon_r \tag{4.10}$$

$$f_{d2} = \frac{1}{2\pi}\frac{\mathrm{d}\phi_2}{\mathrm{d}t} = \frac{1}{2\pi} \times (2\pi\frac{2}{\lambda_2}\upsilon_r) = \frac{2}{\lambda_2}\upsilon_r \tag{4.11}$$

Given the frequency f_1 and f_2 of the radar carrier frequency signal, λ_1 and λ_2 can be calculated. As long as the Doppler shift f_{d1} and f_{d2} are calculated, the target velocity can be obtained.

The intermediate frequency signal of each moving target can demodulate two signals with different phases and close frequencies. The distance of the target can be obtained by analyzing the phase of the two signals. The phase difference of the two signals is

$$\Delta\phi = \phi_1 - \phi_2 = -\frac{4\pi R(f_1 - f_2)}{c} = -\frac{4\pi D\Delta f}{c} \tag{4.12}$$

The distance between radar and moving target is

$$D = -\frac{c\Delta\phi}{4\pi\Delta f}, \quad \Delta\phi \in [0, 2\pi) \tag{4.13}$$

The single target vehicle in the radar irradiation area is shown in Fig. 4.10. The FSK physical radar is installed on the gantry. When the target enters the radar irradiation

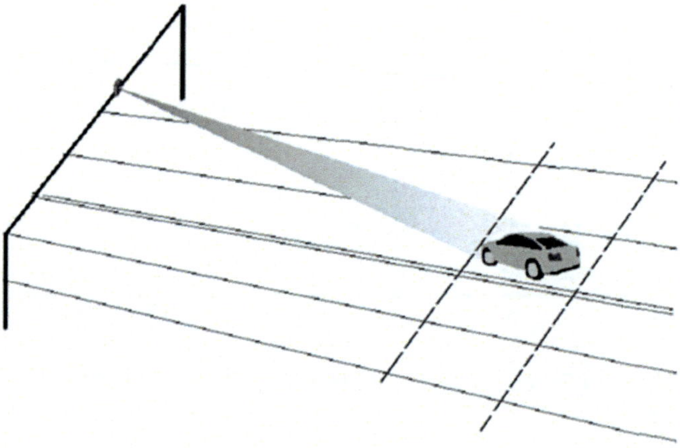

Fig. 4.10 Single target vehicles in radar exposure areas

area, the radar antenna receives the echo signal, and after denoising and spectrum processing, the velocity and distance of the target are obtained.

2. Multi-target recognition of FSK radar

There is a great difference between multi-target and single-target recognition, and their hardware is different. Single-target and double-target recognition are based on a single-antenna microwave radar system, and multi-target recognition is based on a double-antenna microwave radar system. Through the dual-antenna microwave radar system, the three-dimensional information of the target can be measured, including speed, distance, and angle.

The principle of multi-target recognition is the same as that of single-target and binocular identification. The angle measurement principle of multi-target recognition is shown in Fig. 4.11.

At the same time, the frequency of electromagnetic wave emitted by antenna 1 and antenna 2 is the same, the distance between antenna 1 and target is d_1, the distance between antenna 2 and target is d_2, and S is the distance between two antennas, then there are

$$d_3 = d_1 - d_2 \tag{4.14}$$

Since the distance between antenna 1 and antenna 2 is $S << d_1$ and 且 $S << d_2$, the angle between the electromagnetic waves emitted by antenna 1 and antenna 2 is approximately 0, then there are

$$\alpha_1 \approx \alpha_2 \tag{4.15}$$

$$d_3 = S \sin \alpha_2 \approx S \sin \alpha_1 \tag{4.16}$$

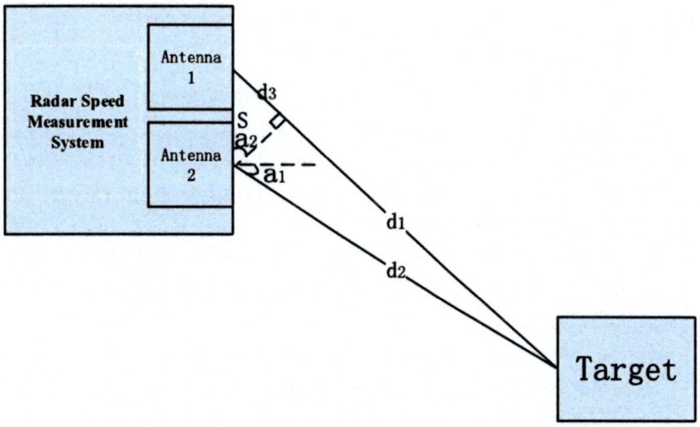

Fig. 4.11 The principle of angle measurement for multi-target recognition

The phase of antenna 1 echo signal is

$$\phi_1 = \frac{2d_1}{\lambda}2\pi \tag{4.17}$$

The phase of antenna 2 echo signal is

$$\phi_2 = \frac{2d_2}{\lambda}2\pi \tag{4.18}$$

From formula (4.14)

$$d_3 = d_1 - d_2 = \frac{(\phi_1 - \phi_2)\lambda}{4\pi} = S\sin\alpha_2 \approx S\sin\alpha_1 \tag{4.19}$$

From formula $\lambda = \frac{c}{f}$

$$d_3 = S\sin\alpha_2 \approx S\sin\alpha_1 = \frac{(\phi_1 - \phi_2)c}{4\pi f} \tag{4.20}$$

where is the frequency of the electromagnetic wave, is the propagation velocity of an electromagnetic wave in the air.

The transverse angle of the target relative to the radar velocity measurement system can be calculated by the formula (4.20). According to the installation height of the radar velocity measurement system and its transverse position relative to each lane, the speed, distance, and angle of the target can be determined.

4.2.2 FSK Radar Target Recognition Algorithm Implementation and Simulation

1. Implementation and simulation of FSK radar single target recognition algorithm

Time domain and frequency domain signals with and without targets are shown in Fig. 4.12 and Fig. 4.13, respectively.

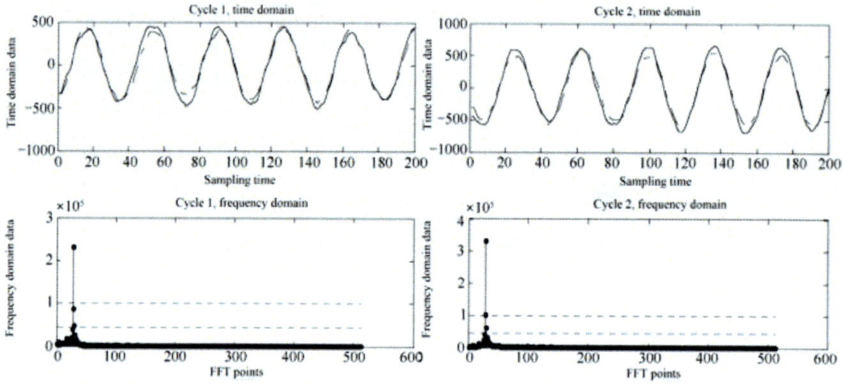

Fig. 4.12 Time domain and frequency domain signals with targets

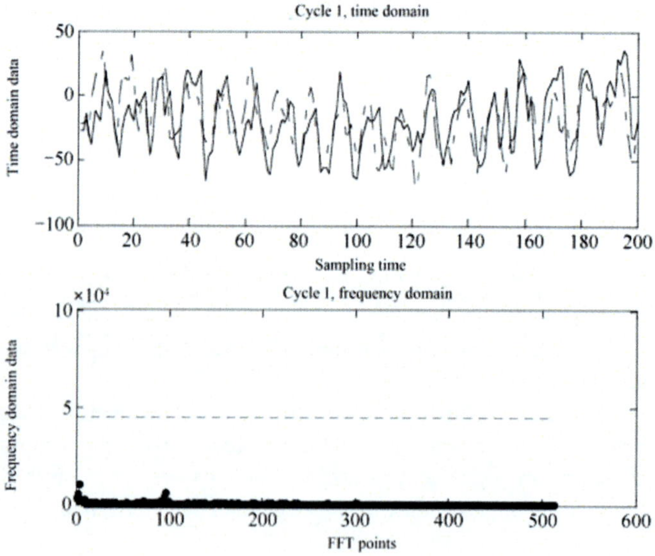

Fig. 4.13 Time domain and frequency domain signals without targets

In Fig. 4.12, the maximum value of the frequency domain signal corresponds to the effective signal of the target, and its frequency is the Doppler shift of the target. It can be seen that FSK radar speed measurement only needs to obtain the frequency and phase of the effective signal. It can be found that in Fig. 4.13, there is no effective signal of the target, and there is still noise residue after denoising. However, the energy of the effective signal of the target is much higher than that of the noise, so it is easy to determine whether there is a target.

The center frequency of the radar used is 24.15 GHz, the sampling frequency is 40 kHz, and the sampling period is 26.24 ms. Therefore, the number of sampling points in each cycle is 1024 points. According to the actual signal and noise characteristics, we designed a set of target recognition and tracking algorithms.

(1) Single target recognition algorithm of FSK radar: threshold thresholds Q_1 and Q_2 (Q_1 is the vehicle arrival threshold and Q_2 is the vehicle departure threshold) are set according to the actual road conditions to determine whether there is a target in the current cycle and whether the target has been driven out of the radar irradiation area. To improve the accuracy of target recognition, the signal-to-noise ratio is used as one of the discriminant conditions. Whether it is for target recognition or target tracking, the signal with the strongest energy should be searched first to obtain the Doppler frequency shift and calculate the moving speed of the target. Signal-to-noise ratio is set. When the amplitude of the frequency point with the strongest energy of the periodic signal is much higher than that of the noise signal and satisfies the signal-to-noise ratio condition, the signal is considered the vehicle target.
(2) Single target tracking algorithm of FSK radar: the range of action of the tracking algorithm is from identifying the target and tracking the target to moving out of the radar irradiation area. Since the signal intensity of the target from entering to leaving the radar irradiation area increases from weak to strong, and then decreases from strong to weak, the vehicle departure threshold Q_2 is lower than the vehicle arrival threshold Q_1. The frequency of the signal is proportional to the velocity, which can be used as the basis for tracking. The radar collects and processes the data every 26.24 ms. In such a short period of time, the velocity of the same target will not change much. Therefore, after identifying the target, the maximum frequency point of each cycle in line with the threshold is judged to determine whether it is the signal of the existing target. If the frequency difference of a continuous multi-cycle signal is large, it is considered that the signal is not the existing target, and the existing target has been driven out of the radar irradiation area. The signal searched in this cycle may be the signal of other targets, which needs to be judged according to the signal-to-noise ratio. In addition, the target distance (the actual distance between target and radar) can be calculated by the phase corresponding to the signal frequency point, and the distance is also the judgment basis of target tracking.

FSK radar single target tracking measured data analysis is shown in Fig. 4.14. Direction 0 indicates that the target is stationary, 1 indicates that the target is heading towards the radar (direction of coming) and −1 indicates that the target is leaving the radar (direction of going).

FSK radar single target recognition and tracking process is shown in Fig. 4.15.

The traditional continuous wave radar can only measure speed. In practical application, the speed radar and the camera constitute the speed radar capture system. When the target enters the radar irradiation area, the radar sends a trigger signal to the camera to trigger the camera to capture the target. Since the traditional radar cannot monitor the target distance, the target position is not controllable when the radar sends the trigger signal. When the target exceeds the speed or violates other traffic rules, it will lead to invalid capture images (such as no license plate, driver, etc.), affecting the traffic management department to confirm the relevant information of illegal vehicles. FSK radar can accurately monitor the target distance, and can send trigger signals at a fixed position to avoid invalid capture.

The comparison of the capture effect between FSK single target radar and ordinary continuous wave radar is shown in Fig. 4.16. It can be seen from Fig. 4.16 that the capture position of FSK single-target radar is relatively stable, while the capture position of ordinary continuous wave radar is very unstable, and even some do not capture the license plate, which affects the effective monitoring of vehicles violating traffic rules by the traffic management department. Statistics show that the effective capture rate of FSK single-target radar is more than 98%, and the fluctuation range of trigger position is within 1 m.

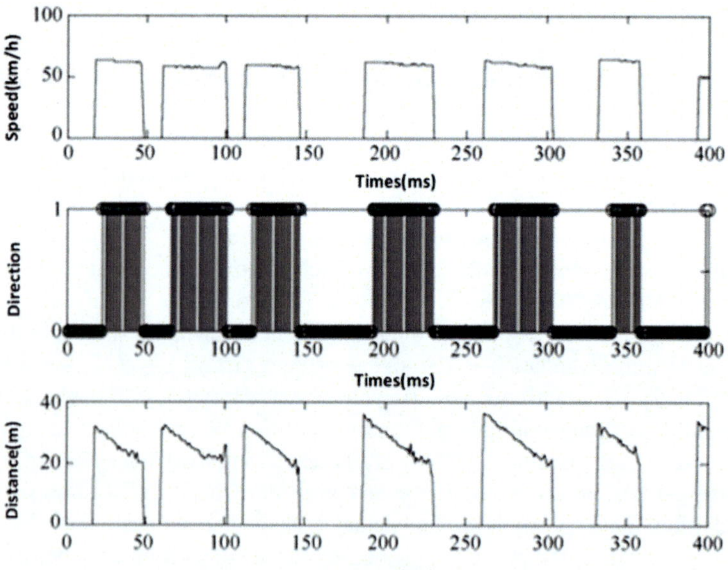

Fig. 4.14 FSK radar single target tracking data analysis

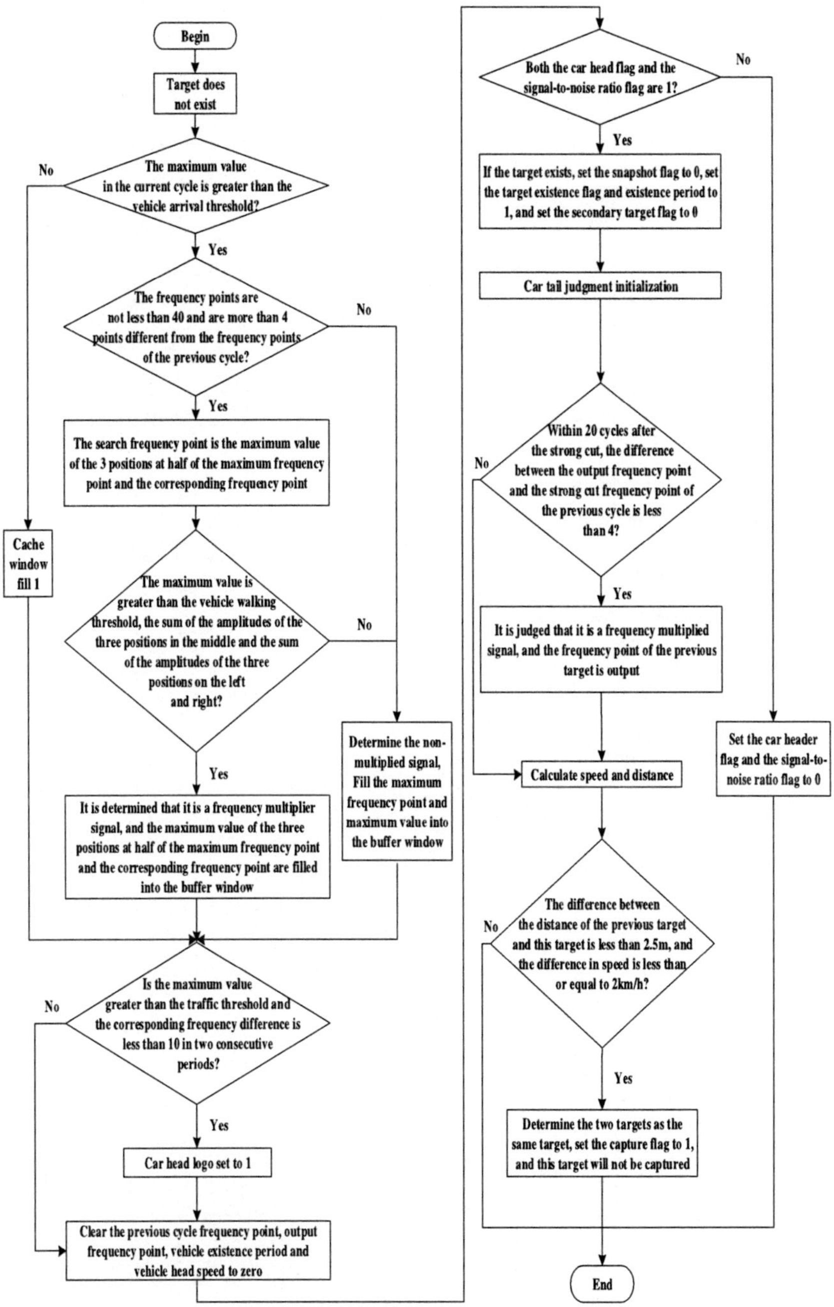

Fig. 4.15 FSK radar single target recognition and tracking process

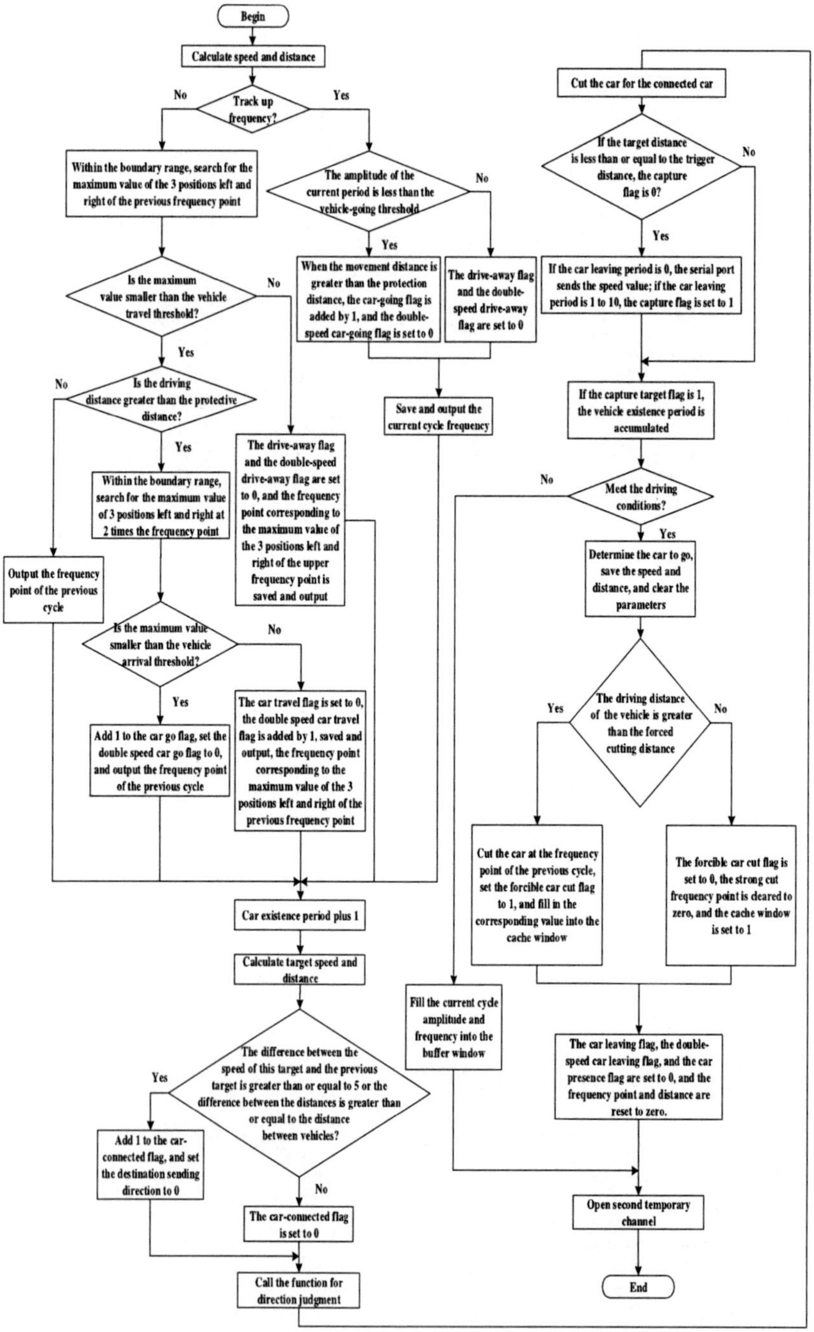

Fig. 4.15 (continued)

(a) FSK single-target radar capture effect

(b) Snapshot effect of ordinary CW radar

Fig. 4.16 FSK single target radar and ordinary continuous wave radar capture effect comparison

2. Implementation and Simulation of FSK Radar Dual Target Recognition Algorithm

This section is based on FSK radar single target speed measurement, in the premise of hardware unchanged, the realization of double target speed measurement. The premise is that the antenna irradiation range can cover two complete lanes. At the same time, two vehicles in different lanes may enter the radar irradiation area, so that a radar can monitor two targets at the same time and reduce the cost.

The time domain and frequency domain signals of the two targets are shown in Fig. 4.17. A single antenna is still used to receive and transmit signals. To identify and track two targets at the same time, the algorithm opens another channel based on one channel, and there is a primary and secondary relationship between the two channels. The tracking mechanism of the main channel is similar to that of the ordinary radar. To reduce the traffic leakage, the recognition and tracking conditions of the target in the secondary channel tracking mechanism are simplified, and the target recognition and tracking of the secondary channel are carried out based on the main channel.

The double target recognition and tracking algorithm of FSK radar are similar to that of a single target. The specific steps are as follows: (1) The search range is determined according to the angle correction value and the upper and lower limits of velocity measurement, and the maximum value and frequency point are searched; (2) Calculate SNR; (3) Determine whether the main channel target vehicle exists, search and track the maximum and frequency points.

In the FSK radar dual target recognition and tracking algorithm, the main target recognition and tracking process can refer to Fig. 4.15, and the secondary target recognition and tracking process are shown in Fig. 4.18.

The measured data analysis of FSK radar dual target tracking is shown in Fig. 4.19.

3. Implementation and Simulation of FSK Radar Multi-target Recognition Algorithm

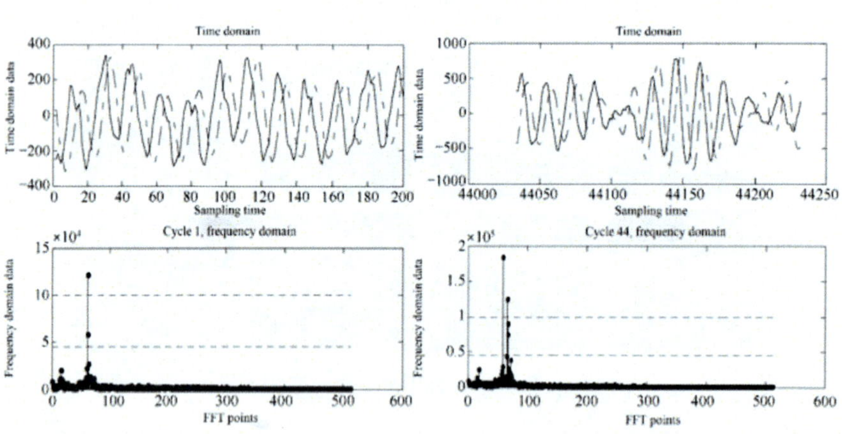

Fig. 4.17 Time domain and frequency domain signals of two targets

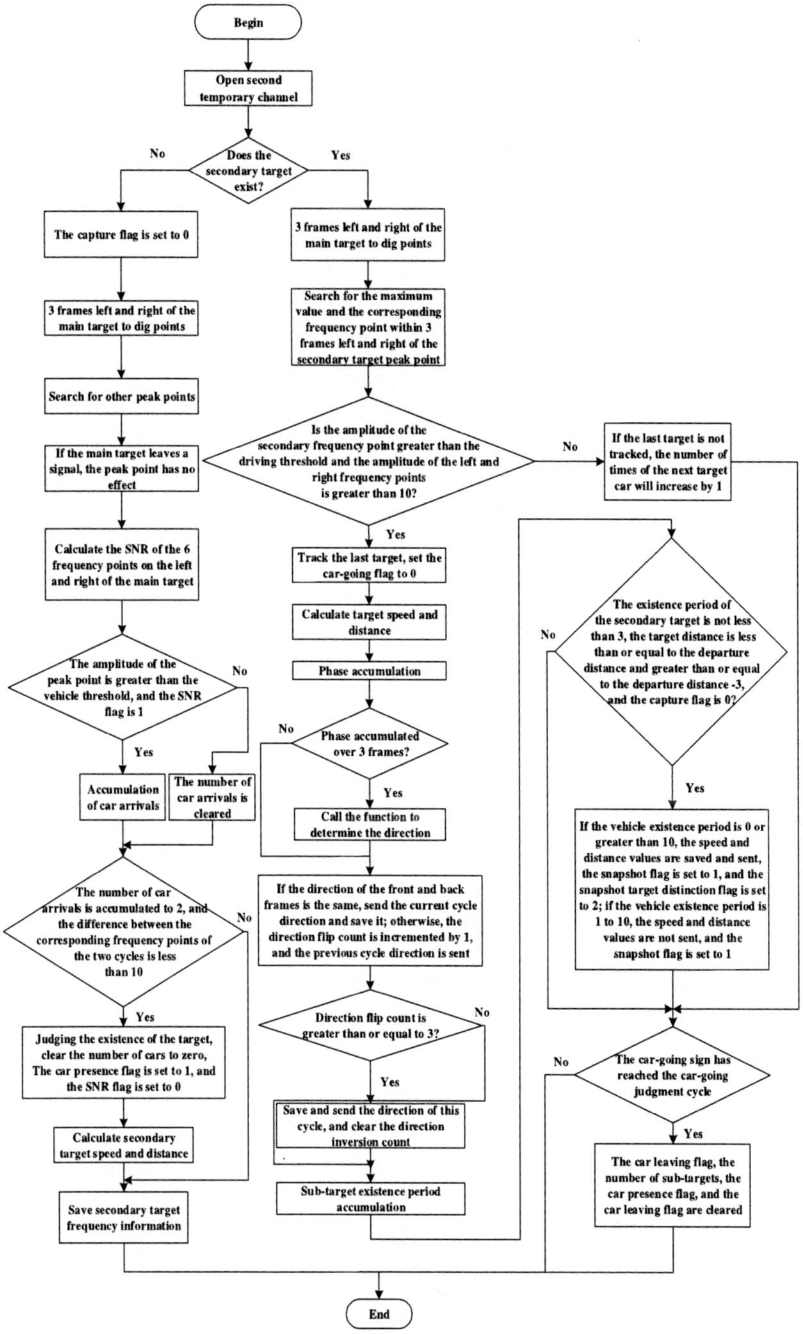

Fig. 4.18 Target recognition and tracking process

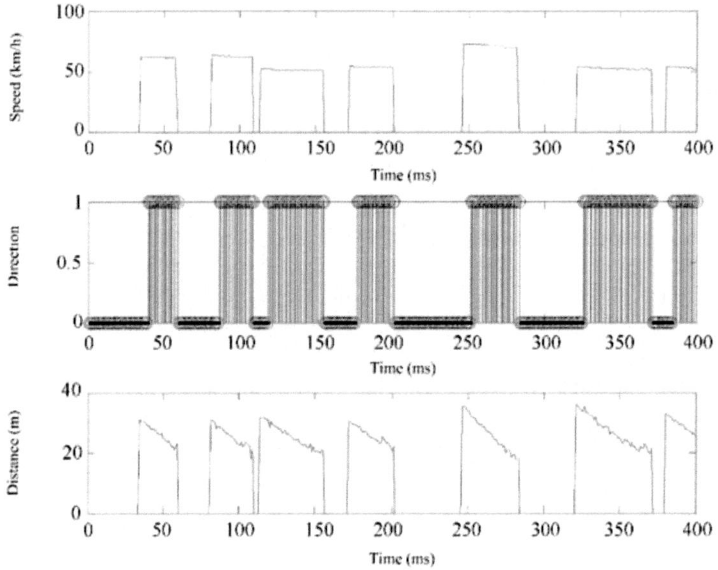

(a) Only one lane with vehicles

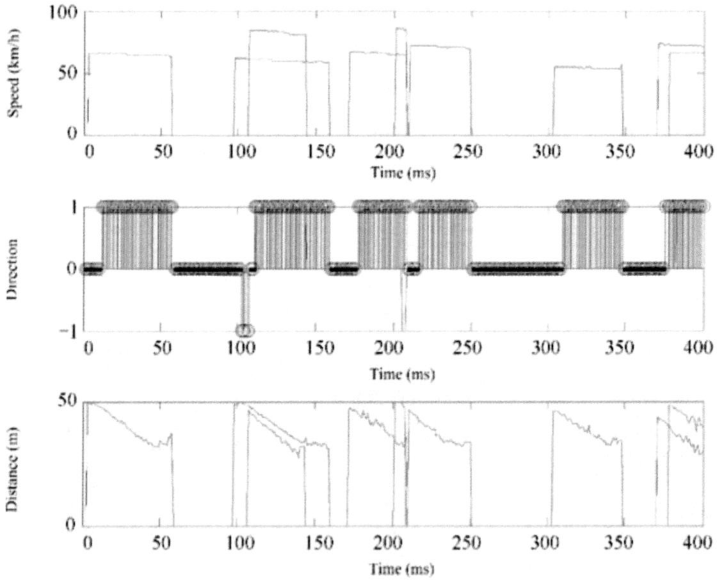

(b) There are cars in both lanes

Fig. 4.19 FSK radar dual target tracking data analysis

The multi-target time and frequency domain signals are shown in Fig. 4.20. Since the complexity of multi-target recognition is much higher than that of single target and double target, it is necessary to make great changes to the multi-target recognition and tracking algorithm. In this algorithm, the radar is installed in the center above the lane.

The specific steps of the algorithm are as follows.

(1) Search for peak points

Since there may be multiple targets in the radar irradiation area, the frequency domain signal obtained by the echo signal after the fast Fourier transform may have multiple peaks. The number of peak points is equal to the number of targets, so the algorithm for searching the maximum frequency point and the sub-maximum frequency point is no longer applicable. To determine the number of targets, the number of effective peak points and the corresponding Doppler frequency shift should be determined first.

(2) Sub-lane

Accurate search for the peak point is the basis of multi-target recognition. According to the Doppler frequency shift corresponding to the peak point, the corresponding distance and transverse angle of each peak point can be calculated, and then the transverse distance of the target relative to the radar can be calculated. The lane can be divided according to the radar installation information.

(3) Calculation of multi-target 3D information

According to the working principle of the radar, the speed, distance, and angle of the target when it enters the radar radiation area are calculated to realize the target positioning.

The above process preliminarily determines that a target enters the radar irradiation area. To avoid the misjudgment of multiple targets in the radar system, it is necessary to conduct real-time tracking of each target before it leaves the radar irradiation area. This process is essentially similar to the case of single target and multi-target. It mainly compares the speed and distance. The difference is that in the case of multi-target, after locking the target lane, it only needs to be compared with the existing target in this lane. If the information is consistent, it is considered that the two are the same targets, that is, the target in the previous cycle in this lane is tracked. If the target in a certain lane is not tracked for three consecutive cycles, it is considered that the target in this lane has been driven out of the radar irradiation area.

(4) Engineering implementation

The proposed FSK radar multi-target recognition and tracking system have been applied in engineering with hardware. The center frequency of radar is 24.15 GHz, the sampling period is 100 kHz, and the number of data points in the frequency domain is 2048 points per cycle. At present, due to the influence of antenna radiation angle, radar can realize multi-target monitoring of three lanes.

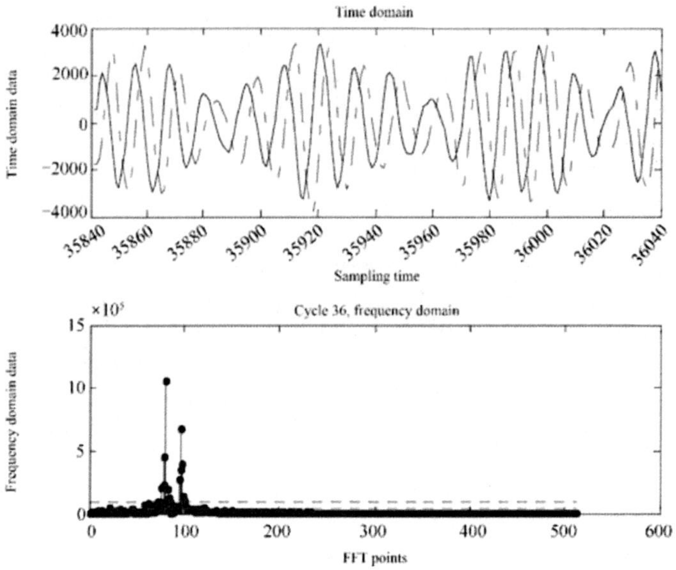

(a) Time domain signal and frequency domain signal corresponding to antenna 1

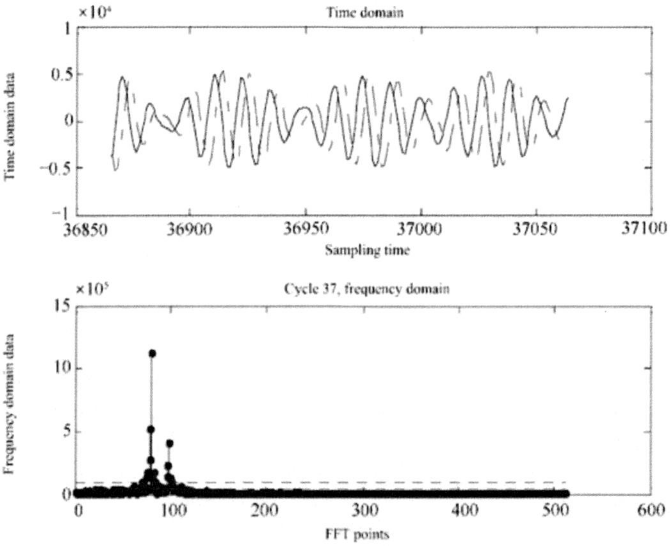

(b) Time domain signal and frequency domain signal corresponding to antenna 2

Fig. 4.20 Time domain and frequency domain signals of multiple targets

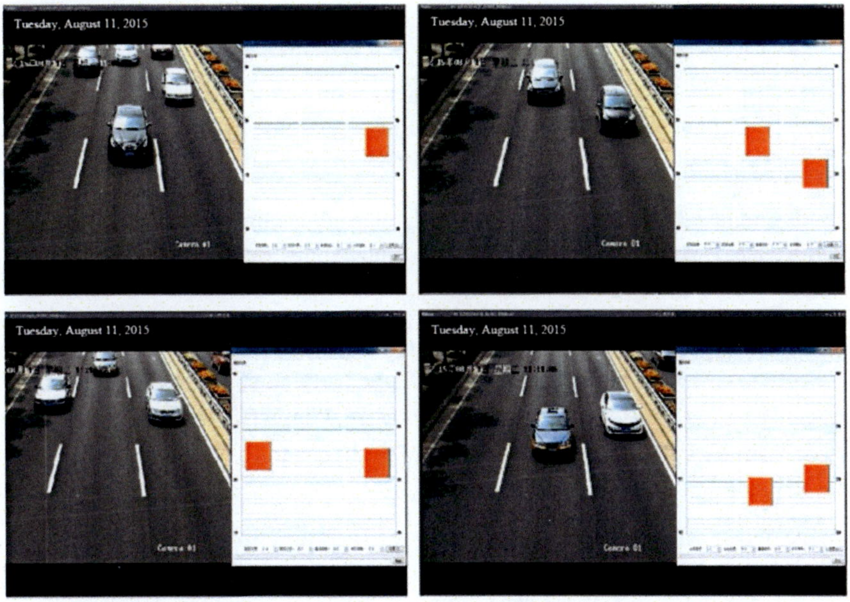

Fig. 4.21 Multi-objective monitoring effect

The multi-target monitoring effect is shown in Fig. 4.21. The left side is the actual running vehicle on the test site, and the right side is the corresponding PC display effect. The small square corresponds to the target of each lane. It can be seen that the radar realizes the multi-target positioning and monitoring.

4.3 Research and Implementation of Multi-target Detection and Tracking Algorithm

The applicability of FSK radar to the multi-target situation is verified. In this section, a multi-target detection and tracking algorithm is proposed based on the FSK system, and the single-frame detection and multi-frame tracking are combined to further improve the detection accuracy by using the correlation between the speed and distance of moving targets.

In the actual installation test, the overall performance of the radar is often reduced due to the pendulum angle error. Based on a large number of collected data, we analyze the data characteristics and extract the feature. Considering the engineering implementation, the self-correction algorithm of radar pendulum angle based on sample statistical characteristics is used to adjust, to improve the target detection rate and scene applicability.

4.3.1 Multi-target Detection and Tracking Algorithm

FSK radar has good anti-noise and anti-attenuation performance, and the spectrum obtained is pure. It can distinguish multiple targets by Doppler frequency shift. Therefore, it is necessary to analyze the spectrum to realize multi-target detection and tracking.

The multi-target detection and tracking algorithm introduced in this section is divided into two parts: single frame detection and multi-frame tracking. Single frame detection mainly completes the elimination and classification of points, eliminates the points that are not satisfied with the conditions, classifies the points, and finds the effective points that fall in each lane; multi-frame tracking determines whether the current target is a false target by multi-frame correlation. In a sense, single frame detection is the basis, multi-and frame tracking is a supplement to single frame detection.

1. Single frame detection

Single frame detection mainly filters targets according to the information of the current frame. In the traffic scene, the vehicle is not a point target, but a multi-scattering point target. Scattering point information fusion is a difficulty, and improper fusion will lead to false alarms or missed alarms. In the radar application scenario described in this book, in most cases, there will be no two vehicles in the same lane laterally, which reduces the difficulty of recognition to some extent.

The spectrum and frequency points in the case of lane separation are shown in Fig. 4.22. There are two main ways of target recognition in the spectrum: one is to find the target first and then divide the lane second, divide the lane first and find the target. When there are targets in multi-lanes at the same time and the speed is close, the spectrum overlap will occur firstly, as shown in Fig. 4.22a. At this time, the spectrum peak is difficult to distinguish and the algorithm is complex, which may lead to vehicle leakage. The second method first divides the target spectrum and then detects the target by searching the maximum value of the corresponding frequency points in each lane range, so that the target is not lost and is easy to implement in engineering. The distribution of the calculated frequency points in the lane is shown in Fig. 4.22c.

To sum up, the single-frame detection is carried out by dividing lanes first and then finding targets. The single-frame detection process is shown in Fig. 4.23.

2. Multi-frame tracking

Single frame detection can obtain the effective frequency points of each lane, but it cannot completely exclude the abnormal clutter in the signal. When a normal vehicle passes through the radar irradiation area, the distance and speed obtained from the original signal have certain rules. The motion characteristic curve of the target is shown in Fig. 4.24. The vehicle detection has the following characteristics.

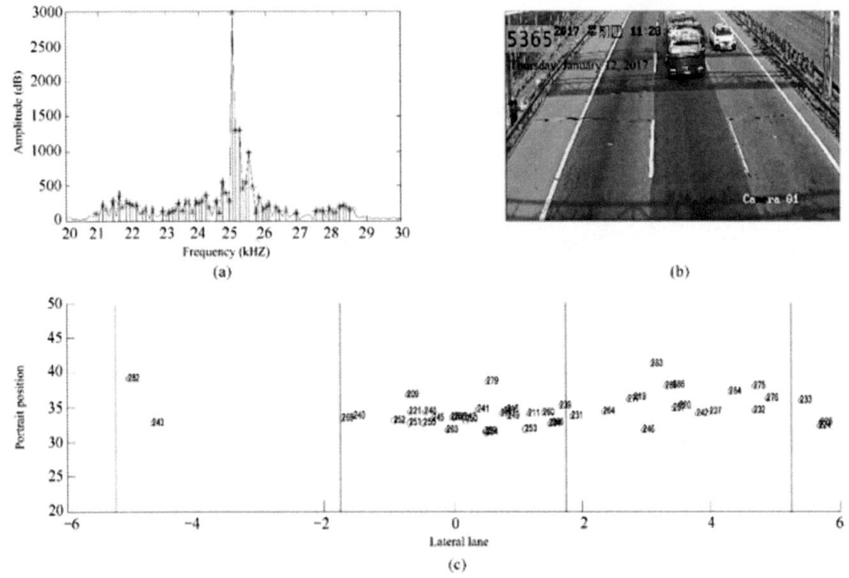

Fig. 4.22 Frequency spectrum and frequency points in lanes

(1) Radar signal changes from weak to strong and then to weak.
(2) Distance decreases linearly.
(3) Speed is basically unchanged.
(4) The target is located in the same lane.
(5) The signal is continuous and stable.

Although the target may do accelerated or decelerated motion, it can be approximately considered to do uniform motion because of its short enough time to pass through the radar irradiation area. If a continuous multi-frame signal of a target approximately conforms to the above five characteristics, it is judged as an effective target.

Based on the above analysis, we propose a speed and distance matching algorithm to achieve multi-frame tracking. The basic idea is that the characteristics of the moving target have a stable correlation, and the distance decreases with time at the current speed.

(1) Measurement of velocity and distance stability

It can be assumed that the speed is constant, so if the speed of the multi-frame signal fluctuates around a certain value, the multi-frame signal may refer to the same target.

Here variance is used to measure the speed stability of multi-frame signals. The discrete point of multi-frame speed is $\{v_{N+1}, v_{N+2}, v_{N+3}, \cdots, v_{N+n}\}$, v_{N+n} is the speed of the current frame, and v_{N+n-1} is the speed of the previous frame.

Average velocity $\overline{v}$ and velocity variance σ_v^2 are obtained

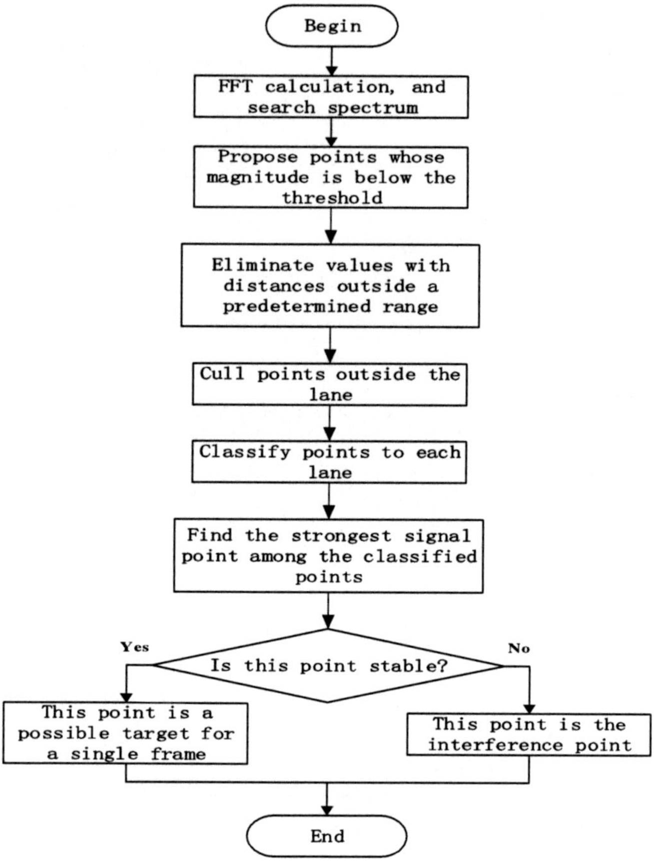

Fig. 4.23 Single frame detection process

$$\overline{v} = \frac{1}{n}\sum_{i=1}^{n} v_{N+i} \tag{4.21}$$

$$\sigma_v^2 = \frac{1}{n-1}\sum_{i=1}^{n} (v_{N+i} - \overline{v})^2 \tag{4.22}$$

The distance decreases linearly with time. If the distance of the multi-frame signal can fluctuate along a straight line, the multi-frame signal may refer to the same target.

In this paper, the stability of multi-frame distance discrete point $\{d_{N+1}, d_{N+2}, d_{N+3}, \cdots, d_{N+n}\}$ is measured by the linear fitting of the least square method with low algorithm complexity commonly used in engineering. d_{N+n} is the distance between the current frame, and d_{N+n-1} is the distance of the previous frame.

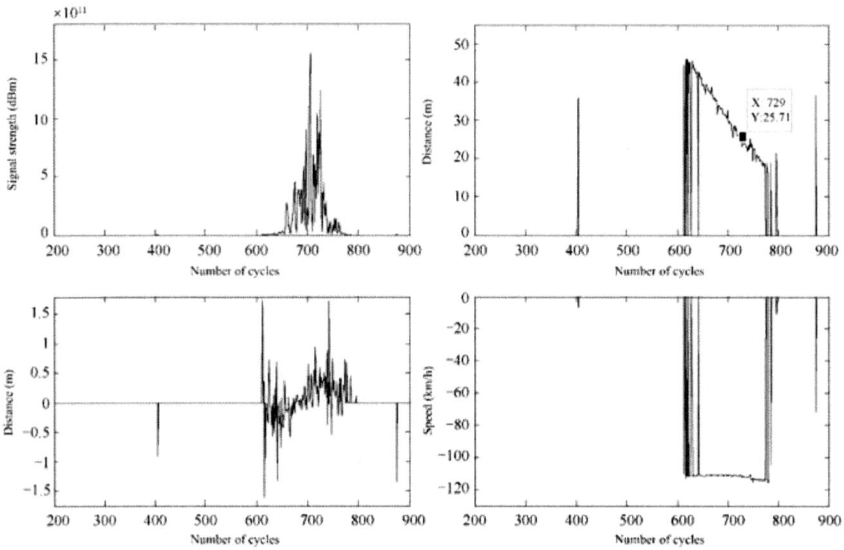

Fig. 4.24 Target motion characteristic curve

The linear fitting model $Y = aX + b$ is taken, where X is the frame number and Y is the corresponding distance of each frame. The slope a and intercept b can be obtained by the least square method as

$$\begin{cases} a = \dfrac{E(XY) - E(X)E(Y)}{E(X^2) - [E(X)]^2} \\ b = \dfrac{E(X^2)E(Y) - E(XY)E(X)}{E(X^2) - [E(X)]^2} \end{cases} \tag{4.23}$$

Distance variance σ_Y^2 is

$$\sigma_Y^2 = \frac{1}{n-1}\sum_{i=1}^{n}(y_i - ax_i - b)^2 \tag{4.24}$$

(2) Measurement of speed and distance matching

The FSK system is similar to the CW system. The target velocity obtained by Doppler frequency shift has high accuracy and small fluctuation, while the target distance is calculated by phase. The existence of clutter will make the distance fluctuate greatly. Therefore, when multiple frames of signals are detected as the same target, the converted distance is obtained according to the speed based on the stable speed information, and the converted distance is matched with the current distance.

Taking the time window of size n, through the speed v_{N+i} obtained by each frame, the slope $a' = v_i T$ is calculated, and the distance obtained by each frame in the time window is converted to the current cycle. The relationship between the converted distance D_i and $\{d_{N+i}\}$ is

$$D_i = d_{N+i} - (n - i)a' \tag{4.25}$$

The distance of all cycles in the time window is converted to the current frame to obtain an array $\{D_1, D_2, \cdots, D_n\}$, whose mean $\overline{D}$ and variance σ_D^2 are

$$\overline{D} = \frac{1}{n}\sum_{i=1}^{n} D_i \tag{4.26}$$

$$\sigma_D^2 = \frac{1}{n-1}\sum_{i=1}^{n} (D_i - \overline{D})^2 \tag{4.27}$$

Mean $\overline{D}$ represents the current distance d_i, and variance σ_D^2 represents the matching degree of speed and distance. It is proved that the speed and distance obtained by this method are stable and the matching degree is high.

The multi-frame tracking process is shown in Fig. 4.25.

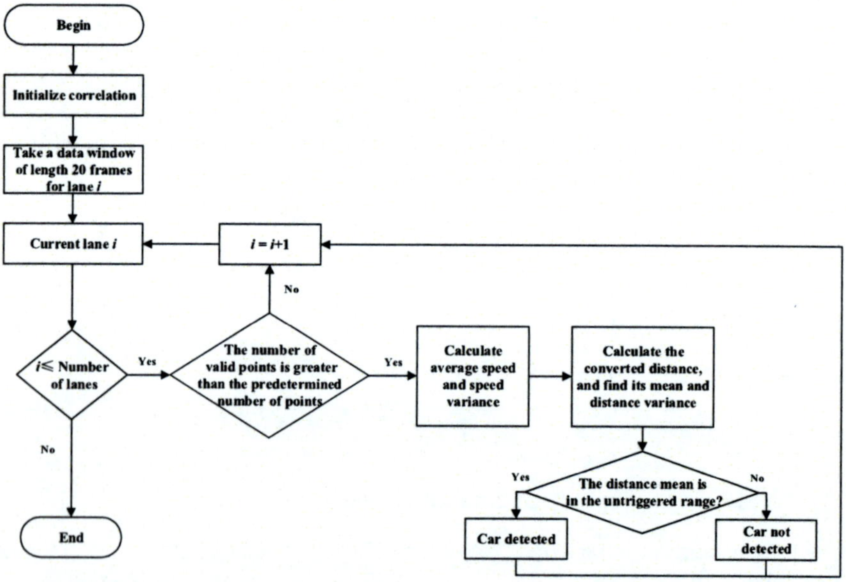

Fig. 4.25 Multi-frame tracking process

4.3.2 Multi-target Tracking Trigger Process

The radar used in this section captures the short-range vehicle at the predetermined position (24 m), and the capture is shown in Fig. 4.26. The vehicle speed and the corresponding lane number need to be given.

The target screening work is completed before, and the target is initially obtained, and the following work needs to be completed later.

(1) Determine whether the target is a detected target or a new target.
(2) Update the detected target information and trigger the capture.

The vehicles in each lane are divided into three categories: triggered vehicles, untriggered vehicles, and untriggered vehicles, which is equivalent to tracking and judging only three vehicles in each lane at the same time. Vehicle classification is shown in Fig. 4.27.

The triggered vehicle refers to the vehicle that has passed the trigger line and is triggered and captured; the vehicle to be triggered refers to the next vehicle to

Fig. 4.26 Capture situation

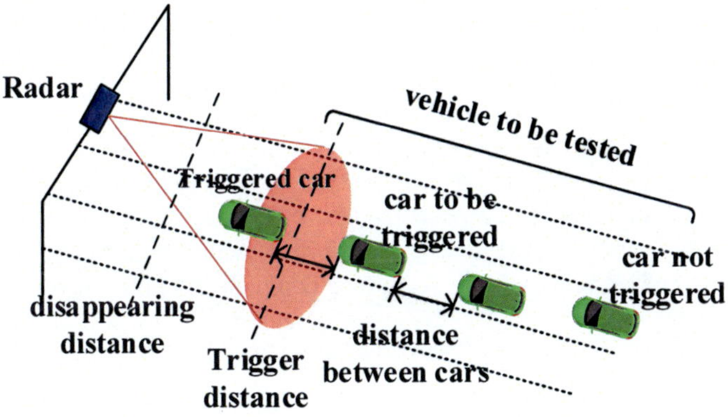

Fig. 4.27 Vehicle classification

pass the trigger line; the untriggered vehicle refers to the second vehicle without a triggering line; the workshop distance refers to the distance between the front and rear cars during driving.

(1) When there is no vehicle in the driveway, the vehicle to be detected shall be classified as the vehicle to be triggered.
(2) When only the triggered vehicle exists in the lane, the vehicle to be detected is compared with the triggered vehicle. If the distance between the vehicle to be detected and the triggered vehicle is less than the minimum workshop distance, the vehicle to be detected and the triggered vehicle are matched. The information of the triggered vehicle is updated with the information of the vehicle to be detected, and it is no longer tracked after exceeding the vanishing distance. If the distance between the vehicle to be detected and the triggered vehicle is greater than or equal to the minimum workshop distance, the vehicle to be detected is classified as the vehicle to be triggered.
(3) When there is an untriggered vehicle in the lane, the detected vehicle is compared with the untriggered vehicle. If the distance between the vehicle to be detected and the vehicle to be triggered is less than a certain value, the vehicle to be detected matches the vehicle to be triggered, updates the vehicle to be triggered information with the vehicle to be detected information, and becomes the vehicle to be triggered after exceeding the trigger distance; if the distance between the vehicle to be detected and the vehicle to be triggered is greater than or equal to that value, the vehicle to be detected is classified as an untriggered vehicle.
(4) When there is an untriggered vehicle in the lane, the vehicle to be detected is regarded as invalid information and discarded.

The above process is a fusion process of information matching and updating. For a complete tracking process, the self-renewal process of target information is indispensable. In most cases, the target state (distance, speed) is determined and does not change in a short time. Therefore, in the self-renewal process of target information, consider using the following formula

$$d_n = d_{n-1} - v\Delta t \tag{4.28}$$

where A is the distance of the target in period B, C is the speed of the target, and D is the duration of each cycle.

When each target has matching information, match updates, otherwise self-updates.

Target tracking constantly repeats the above process. The untriggered vehicle is introduced into the vehicle to be detected, and the vehicle state transition continuously makes the triggered vehicle leave, forming a complete tracking trigger process.

4.3.3 Optimization of Target Detection Based on Constant False Alarm Detection

1. Constant false alarm detection

When the radar is applied to the road traffic environment, the radar echo received by the antenna includes not only the echo signal of the moving target, but also the clutter signal formed by the scattering of the well cover and the isolation belt on the road and the interference signal such as system noise. Therefore, when the radar detects the target from the echo signal, the detection threshold is usually set. When the signal is higher than the threshold, the target is judged. When the signal is below the threshold, the target is not considered to exist. When the threshold setting is too low, clutter will be detected as a target, resulting in a false alarm; when the threshold is too high, the target may not be detected, resulting in missed alarm.

It can be seen that the selection of threshold is crucial, and the probability of false alarm and the missed alarm is usually used to measure the performance of radar detection. A radar with good detection performance requires reducing the miss rate as much as possible based on not exceeding the expected false alarm rate.

The radar detection threshold is usually related to clutter and noise. In practical traffic applications, clutter, noise, and signal strength of the target in echo signal are constantly changing. Therefore, it is necessary to adopt a dynamic threshold to adapt to the change in echo signal.

Constant false alarm detection obtains an adaptive dynamic threshold according to the characteristics of the signal, which keeps the false alarm rate of radar unchanged and improves the detection performance of the radar.

Common constant false alarm detection methods include unit average constant false alarm detection, ordered statistics constant false alarm detection, etc.

(1) Unit average constant false alarm detection (CA-CFAR)

Each L unit $\{x_{i-L}, \ldots, x_{i-1}, x_i, x_{i+1}, \ldots, x_{i+L}\}$ around the input signal x_i is calculated, and the estimation $\hat{\mu}$ of the clutter background is

$$\hat{\mu}=\frac{1}{2L+1}\sum_{n=-L}^{L} x_{i+n} \tag{4.29}$$

Define $\hat{\mu}$ as the basic threshold.

At this time, the radar detection threshold U_0 can be expressed as

$$U_0 = K\hat{\mu} \tag{4.30}$$

Equation (4.30) formula, K is the threshold multiplier factor to adjust the false alarm rate.

It can be seen that this method takes the finite element to estimate the clutter background, and the $\hat{\mu}$ obtained at different times will fluctuate. The less the number of elements L is, the greater the fluctuation of $\hat{\mu}$ is. From the practical application, the value of L cannot be infinite. Therefore, this method is suitable for the average clutter environment, and its effect will deteriorate in a non-uniform clutter environment.

(2) Ordered statistical CFAR detection (OS-CFAR)

Each L element $\{x_{i-L}, \ldots, x_{i-1}, x_i, x_{i+1}, \ldots, x_{i+L}\}$ of the input signal x_i is sorted so that the sorted array $\{X_1, X_2, X_3, X_4, \ldots, X_{2L+1}\}$ satisfies $X_1 \leq X_2 \leq X_3 \leq \cdots \leq X_{2L+1}$, and the mth element X_m is used as the estimation of the clutter level, that is, the basic threshold is set as

$$\hat{\mu}=X_m \tag{4.31}$$

At this time, the radar detection threshold U_0 can be expressed as

$$U_0=K\hat{\mu}=KX_m \tag{4.32}$$

It can be seen that this method takes the first unit in the sorted finite element for clutter background estimation. When there is strong clutter interference in the unit near the input signal at a certain time, although it will have an impact on the sorting of the unit, it is almost not affected. Therefore, this method has a strong anti-interference ability. In the non-uniform clutter environment, it has a strong detection ability and is suitable for multi-target scenarios.

To further improve the anti-interference ability of OS-CFAR, the following improvements are made.

(1) Delete and average detection (CMLD-CFAR), exclude several maximum points, and calculate the average value of the remaining data as the basic threshold.
(2) Organize and average detection (TM-CFAR), exclude several maximum points and minimum points, and calculate the average value of the remaining data as the basic threshold.

2. Application of constant false alarm detection

The target signal that the radar scans at some time is shown in Fig. 4.28. In Fig. 4.28, the tadpole-like dark color signal is the target signal, and its clutter is. The stronger the target signal is, the deeper the color is.

It can be seen from Fig. 4.28 that the stronger the target signal is, the stronger the clutter is (the more obvious the texture is), so the constant false alarm detection should be used to avoid false alarms.

Taking the sorted median as the basic threshold μ_1, the detection threshold is $U_0' = K\mu_1$, which is used to detect the signal. The detection effect of the constant false alarm dynamic threshold is shown in Fig. 4.29b.

To compare, the low threshold fixed threshold and high threshold fixed threshold are used to process the signal, and the detection results are shown in Fig. 4.29c, d, respectively. The original image contains all clutter and target signals, as shown in Fig. 4.29a. The stronger the signal is, the deeper the color is.

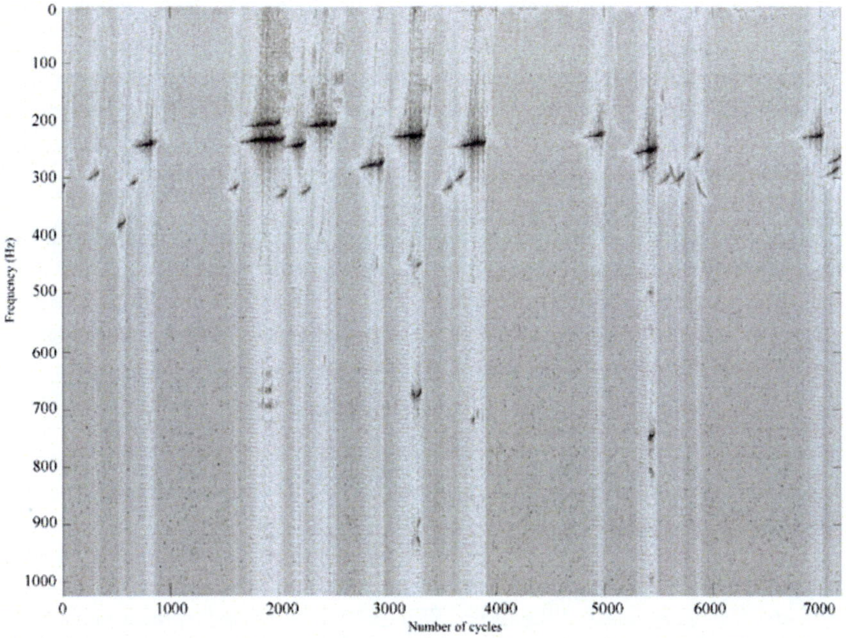

Fig. 4.28 The target signal scanned by radar at some time

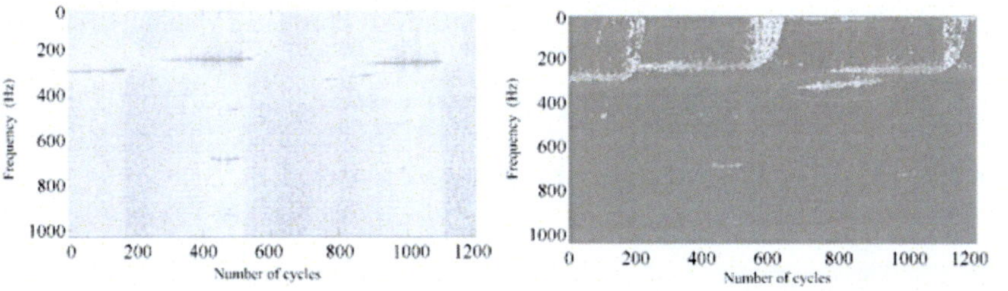

(a) The original image

(b) The detection effect of using the constant false alarm dynamic threshold

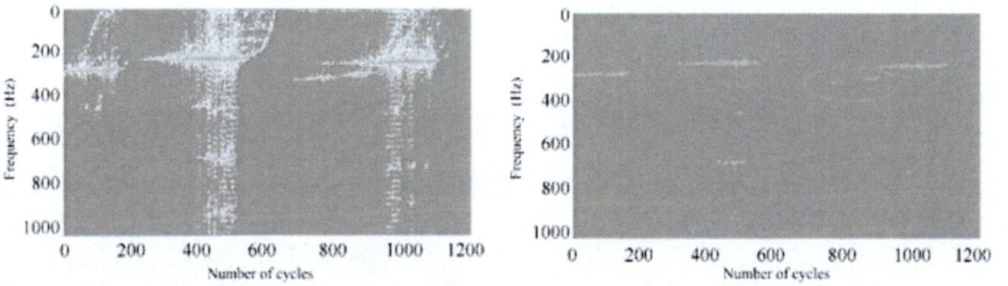

(c) Detection effect with a low-threshold fixed threshold

(d) Detection effect with a high-threshold fixed threshold

Fig. 4.29 Detection effect

It can be seen from Fig. 4.29 that when the low threshold fixed threshold is used to detect the target, many clutter signals are not filtered, which may cause a false alarm. When the high threshold fixed threshold is used to detect the target, the clutter signal has no residual, but a part of the weak target signal is filtered, which may cause missed alarm; when the target is detected by a constant false alarm dynamic threshold, not only the clutter can be well suppressed, but also the weak signal detection ability can be realized, and the weak signal has strong robustness.

The constant false alarm detection uses an adaptive threshold instead of a fixed threshold, which reduces the requirements of manual intervention. In the actual industrial production debugging process, constant false alarm detection reduces the amplitude difference caused by the individual differences of scene and radar, and enhances the performance.

3. Improvements in the engineering implementation

In Fig. 4.29b, ordered statistical CFAR detection is used to sort the whole spectrum and estimate the noise of the whole spectrum. The calculation is large and needs to be improved in engineering implementation.

The signal amplitude is shown in Fig. 4.30. It can be seen that the target signal only accounts for a small part of the spectrum, most of which are noise and relatively uniform. This is because of the use of FSK radar and applied to short-range traffic scenarios (irradiation range is not more than 70 m), the target speed is relatively close.

The order of multi-period signal amplitude is shown in Fig. 4.31. It can be seen from Fig. 4.31 that the strength of clutter is positively correlated with the strength of the target signal.

Based on the above analysis, it is considered to select a point without a target in the spectrum and use this point to estimate the basic threshold of the signal. Based on this consideration, two simplified methods are proposed.

Fig. 4.30 Signal amplitude

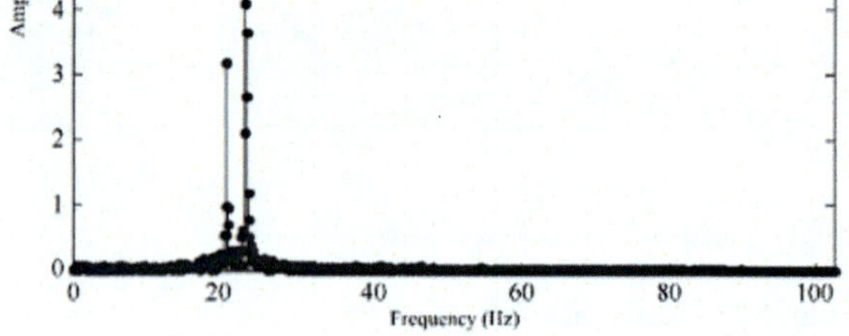

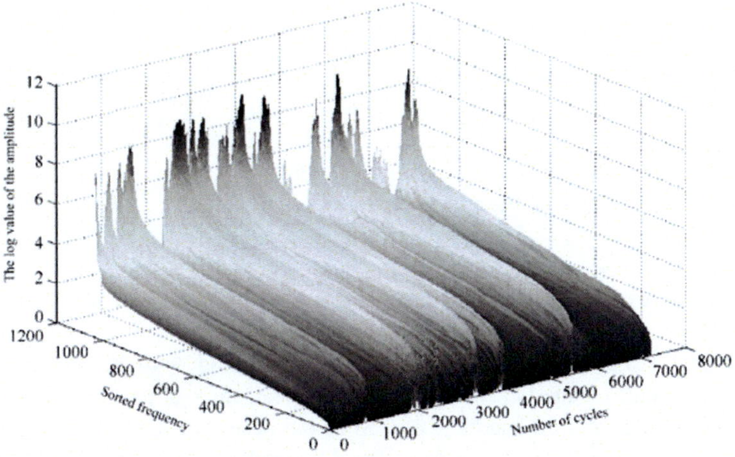

Fig. 4.31 Multi-period signal amplitude sorting

(1) Division average sorting method

The whole spectrum is divided into N intervals $\{[x_1, \cdots, x_{i_1}], \cdots, [x_{N_1}, \cdots, x_{Ni_1}]\}$, the mean value of N intervals is calculated respectively, and the N means unit $\{\mu_1, \mu_2, \cdots, \mu_N\}$ is obtained. The k mean unit μ_k is used as the basic threshold.

$$\hat{\mu} = \mu_k = \frac{1}{N}\sum_{i=1}^{N}\mu_i \tag{4.33}$$

The detection threshold is

$$U_0 = K\hat{\mu} = K\mu_k \tag{4.34}$$

(2) High-frequency screening point method

The velocity resolution of the radar used is 0.34 km/h. It is estimated from the spectrum that the maximum velocity that the radar can measure can reach 300 km/h, and the driving speed of the vehicle usually does not exceed 200 km/h. It can therefore be considered that there are no targets in the high-frequency region. At this point, the basic threshold can be obtained by estimating the clutter distribution of the high-frequency spectrum.

In the high-frequency region, each unit $\{x_{f_n-L}, \cdots, x_{f_n-1}, x_{f_n}, x_{f_n+1}, \cdots, x_{f_n+L}\}$ of L around signal x_{f_n} is taken and sorted to make the sorted array satisfy $X'_1 \le X'_2 \le X'_3 \le \cdots \le X'_{2L+1}$. The mth unit X'_m is used as the estimation of clutter level in the high-frequency region, that is, the basic threshold of the whole spectrum clutter background is

$$\hat{\mu} = X'_m \tag{4.35}$$

The detection threshold is

$$U_0 = K\hat{\mu} = KX'_m \tag{4.36}$$

This method draws on the sorting and averaging method in ordered statistics.

4. Test results

The basic threshold curves of signals obtained by different methods are shown in Fig. 4.32.

Figure 4.32a use the partition average sorting method, with 50 intervals and 20 frequency points covered in each interval. The mean value of each interval is calculated and sorted, and the median is taken as the basic threshold of a single frame signal, to obtain the basic threshold curve of the multi-frame signal.

Figure 4.32b Sorting 800–900 points in the high-frequency region and averaging 51 points in the middle of the sorting as the basic threshold of single frame signal, the basic threshold curve of the multi-frame signal is obtained.

Figure 4.32c ranks the full spectrum, and takes the sorted median as the basic threshold of a single frame signal, thus obtaining the global optimal basic threshold curve.

Taking the global optimal basic threshold curve in Fig. 4.32c as the contrast, the correlation between the results obtained by the improved two methods and their correlation is calculated respectively. The correlation coefficient of the partition average

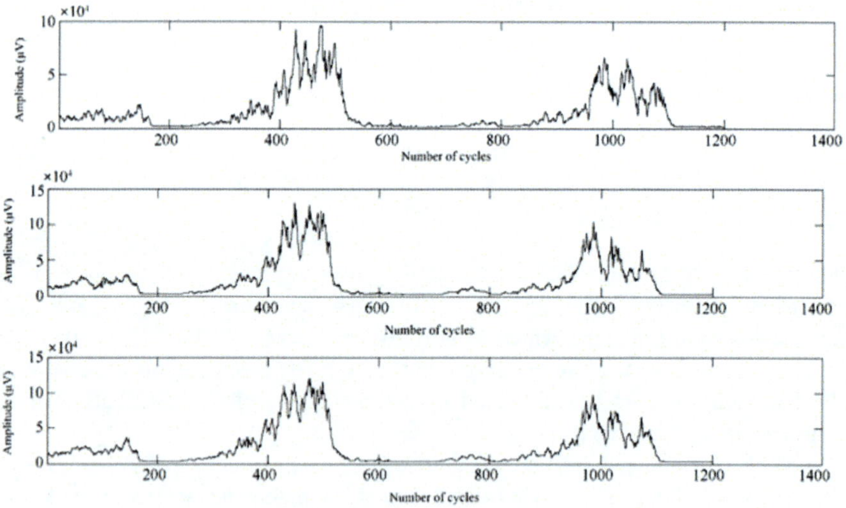

Fig. 4.32 Basic signal threshold curves obtained by different methods

sorting method of Fig. 4.32a is 0.99471, and the correlation coefficient of the high-frequency screening point method of Fig. 4.32b is 0.98781. The closer the correlation coefficient is to 1, the stronger the correlation is.

It can be seen that the two selected methods have a significant indigenous correlation with the optimal threshold, and the correlation in Fig. 4.32a is stronger, indicating that the improved two methods greatly reduce the calculation amount without much performance degradation. Moreover, the performance of the partition average sorting method is better than that of the high-frequency screening point method, and its calculation amount is also smaller than that of the high-frequency screening point method.

4.3.4 Test and Analysis

The above optimization algorithm is written into the DSP module to test, observe the actual work of the system, count the effective capture rate, collect data for MATLAB simulation, compare with the test situation, verify the feasibility of the algorithm, and constantly optimize according to the actual situation.

1. Multi-lane testing

The test scene is a highway, measuring three lanes in one direction, and the multi-lane test scene is shown in Fig. 4.33.

The interrupt period of the radar is about 8 ms, and 256 points of each signal are collected in each interrupt period. The time-domain signal and frequency-domain signal of a single lane are shown in Fig. 4.34. The maximum signal strength corresponds to the Doppler frequency shift of the target, and the velocity of the target can be obtained by this value. The time-domain signal and frequency-domain signal of the three lanes are shown in Fig. 4.35. This situation is relatively complex. It can be seen that the frequency-domain signal has three peaks, representing the Doppler frequency shift of the three targets, respectively. Thus, the velocity of each target can

Fig. 4.33 Multi-lane test scenario

be obtained. To further distinguish the target distance and lane number information, it is necessary to calculate the phase difference of the relevant signal.

The information obtained from the simulation of the measured data is shown in Fig. 4.36. In Fig. 4.36, the three envelope waveforms with the period number of about 1200 are the situation that there are three lanes in Fig. 4.35. Figure 4.36 clearly show

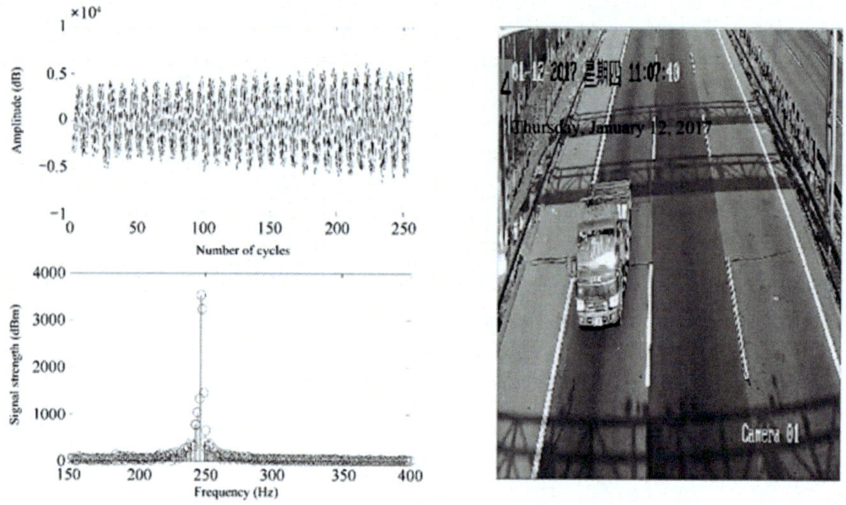

Fig. 4.34 Time domain signal and frequency domain signal of the single lane with a vehicle

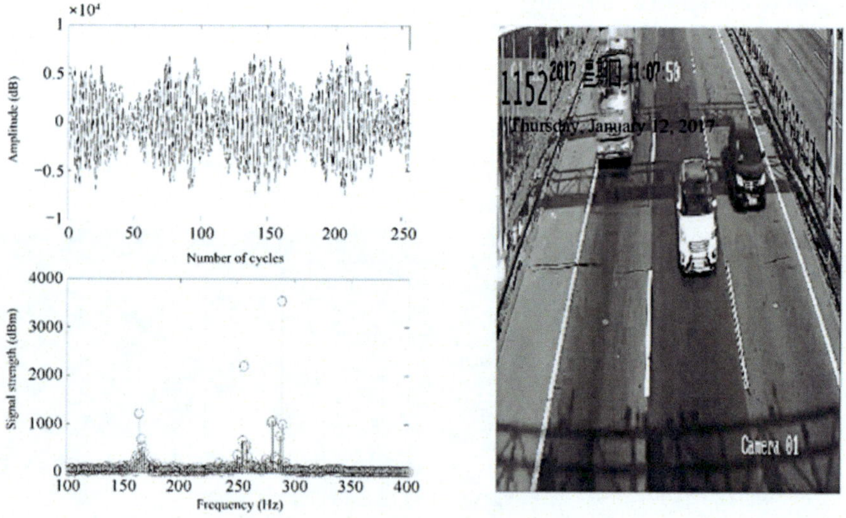

Fig. 4.35 Time domain and frequency domain signals when three lanes have cars at the same time

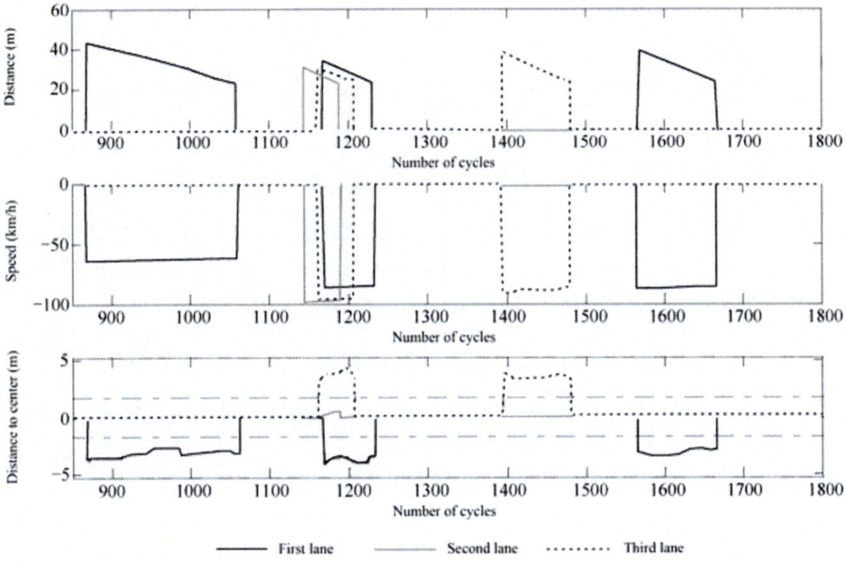

Fig. 4.36 Information obtained by simulating the measured data

the distance, speed, and distance to the center of the target. It can be verified that the algorithm can simultaneously monitor multiple targets in multiple lanes.

2. Effective capture rate

The effective capture rate of the radar in the multi-target scene is shown in Table 4.2 by comparing the captured video with the captured image.

It can be seen from Table 4.2 that the effective capture rate is more than 95%, which is greatly improved compared with 90% of CW radar.

Table 4.2 Effective capture rate of the radar in the multi-target scene

	Group 1	Group 2	Group 3	Group 4
Number of shots	316	334	285	301
Missing shots	13	15	11	9
Side-by-side missed shots	11	14	9	9
Multiple shots	8	12	7	5
The number of wrong lanes	3	5	2	2
Number of valid shots	308	322	278	296
Actual number of vehicles passed	321	337	289	305
Effective capture rate (%)	95.95	95.55	96.19	97.04
Missed beat rate (%)	4.05	4.45	3.81	2.96
Multi-beat rate (%)	2.49	3.56	2.42	1.64

Fig. 4.37 FSK radar capture effect

Since such products are still in the research and development stage and involve commercial secrets, the theoretical research of radar algorithms in multi-target scenes has not seen public literature and practical engineering application effect, so it is only compared with the existing standards. The traffic management department requires that the capture rate of velocity radar is higher than 90%, so the proposed algorithm can meet the actual needs of road traffic.

3. Capture effect

FSK radar ranging is accurate, can set the radar at a fixed distance trigger, FSK radar capture effect as shown in Fig. 4.37.

It can be seen from Fig. 4.37 that the target motion information can be accurately obtained in the case of a single lane and multi-lane at the same time. No matter the size of the target, the type of vehicle, and the location of the lane, the capture is almost triggered at the same lateral position.

4.4 Summary

This chapter introduces the FSK system one-dimensional, two-dimensional, three-dimensional radar algorithm implementation process, proposed and implemented a multi-target radar algorithm, covering the recognition algorithm, and tracking algorithm, through the actual road test verification and analysis. This chapter proposes a multi-target detection and tracking algorithm based on the FSK system. To further improve the accuracy of target detection, single frame detection, and multi-frame tracking are combined. To improve the false alarm and missed alarm phenomena that are easy to occur in multi-target detection, the ordered statistical constant false alarm detection method is adopted, and the detection threshold is set. Combined with the engineering implementation, the improved method is proposed, and the performance is compared with the optimal threshold. In addition, combined with the actual application of the radar swing angle problem, this chapter completes the feature extraction based on the analysis of a large number of data characteristics, and puts forward the self-correction algorithm of radar swing angle based on the statistical characteristics of samples, which reduces the capture error caused by swing angle. The actual test of the radar system is carried out, and multiple targets in multiple lanes are monitored. The effective capture rate reaches 95% and has a good capture effect.

Chapter 5
Clustering Algorithms

The advantages of the clustering algorithm and the classical clustering algorithm are discussed in this chapter. For the problem of distinguishing adjacent objects in the highway traffic scene, the classical clustering algorithm is improved, and the main axis-based density peak fuzzy clustering (SDPFC) algorithm is proposed. This chapter tests and simulates the actual scene compares the SDPFC clustering algorithm with the DBSCAN clustering algorithm, the FCM clustering algorithm, and the K-Means clustering algorithm, and verifies the superior performance of the SDPFC clustering algorithm.

5.1 Classical Clustering Algorithms

The sample points obtained by signal preprocessing are scattered and accompanied by noise. Therefore, before multi-target tracking, clustering sample points can better complete multi-target tracking. Several common clustering algorithms are as follows.

5.1.1 DBSCAN Clustering Algorithm

The DBSCAN clustering algorithm is a spatial clustering algorithm, which is classified according to the density characteristics of the clustered data. The clustering result of this algorithm is the largest cluster of density-connected sample points, the quality of the clustering result is mainly related to the neighborhood radius σ and the selection of the minimum number (minpts) of sample points within the neighborhood radius.

Neighborhood σ: σ represents the radius of a circle centered on a sample point. The neighborhood sample point set D of this sample point is defined as other sample

L. Cao et al., *Target Recognition and Tracking for Millimeter Wave Radar in Intelligent Transportation*, https://doi.org/10.1007/978-981-99-1533-0_5

points in the circle centered on a sample point and with a radius σ in the candidate set N, which is as follows:

$$N\sigma(z_i) = \{z_j \in D | \mathrm{dist}(z_i, z_j) \leq \sigma\} \tag{5.1}$$

Core sample point: Any sample point $z_i \in D$, if there are at least minpts sample points in its neighborhood, z_i is called the core sample point, which is expressed as

$$|N\sigma(z_i)| \geq \text{minpts} \tag{5.2}$$

Directly density-reachable: If the sample point z_i is the cluster center, then all the sample points z_j in its neighborhood sample point set D are directly density-reachable from z_i.

Density-reachable: If there is an object chain $p_1, p_2, \cdots, p_T$, which satisfies $p_1 = z_j$, $p_T = z_i$, and p_{k+1} is directly density-reachable from p_k, thus z_j is density-reachable from z_i.

Density-connected: If there is a point $z_k \in D$, both z_j and z_i are density-reachable from z_k, then z_j is density-connected to z_i.

To better understand the parameters defined by the DBSCAN clustering algorithm, a diagram of the DBSCAN clustering algorithm is given as shown in Fig. 5.1. The white points are core sample points, and the black points are non-core sample points. As can be seen from Fig. 5.1, the minpts is 5 because there are at least 5 sample points within σ-neighborhood of the core sample point. All the directly density-reachable sample points are in the circle centered on the core sample point, and the arrows form the density-reachable sample point sequence. In the σ-neighborhood of these density-reachable sample points, all the sample points are density-connected.

According to the parameters of the DBSCAN clustering algorithm, the maximum set of density-connected sample points is the clustering result of the algorithm, and each category in the clustering result can have one or more core sample points. If there is only one core sample point in a category, the other core sample points in the category are all within the σ-neighborhood of this core sample point; if there

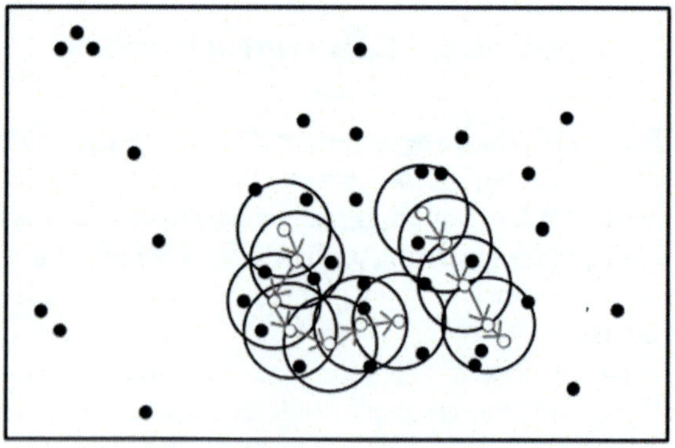

Fig. 5.1 Diagram of the DBSCAN clustering algorithm

are multiple core sample points in a category, Then there is at least one other core sample point in the σ-neighborhood of any core sample point in this category; the set of all sample points in the σ-neighborhood of these core sample points constitutes a DBSCAN clustering category.

The process of the DBSCAN clustering algorithm is shown in Fig. 5.2.

The DBSCAN clustering algorithm can cluster sample point sets with arbitrary shapes, which is very suitable for traffic scenes. However, the algorithm needs to jointly adjust the parameters σ and minpts find the optimal solution in the current scene which is quite complicated for users. In addition, because the sample points of adjacent vehicles are close and the density is similar, the DBSCAN clustering algorithm cannot distinguish adjacent vehicles in traffic scenes well.

5.1.2 FCM Clustering Algorithm

The FCM clustering algorithm is a fuzzy computational clustering algorithm obtained by hard clustering. The algorithm uses the objective function to calculate the fuzzy index. The clustering result does not belong to or does not belong to this binary result, but is represented by a numerical value of 0 ~ 1, which is called membership degree. Furthermore, the FCM clustering algorithm is automatic and requires no human intervention.

Suppose that the sample point is $z_j(j = 1, 2, \cdots, k)$, if these data are divided into i categories, then the corresponding center is i_c. Define the objective function as J. The membership degree of the sample point belonging to a certain category is donated as μ_{ij}. Then the expression of the objective function J is as follows:

$$J = \sum_{i=1}^{I} \sum_{j=1}^{k} \mu_{ij}^{m} \left\| z_j - i_c \right\|^2 \tag{5.3}$$

For each sample point z_j, the sum of membership degrees belonging to the class i should be 1, then there are

$$\sum_{i=1}^{I} \mu_{ij} = 1, j = 1, 2, \cdots, k \tag{5.4}$$

According to the properties of the objective function, it can be extended to get the following:

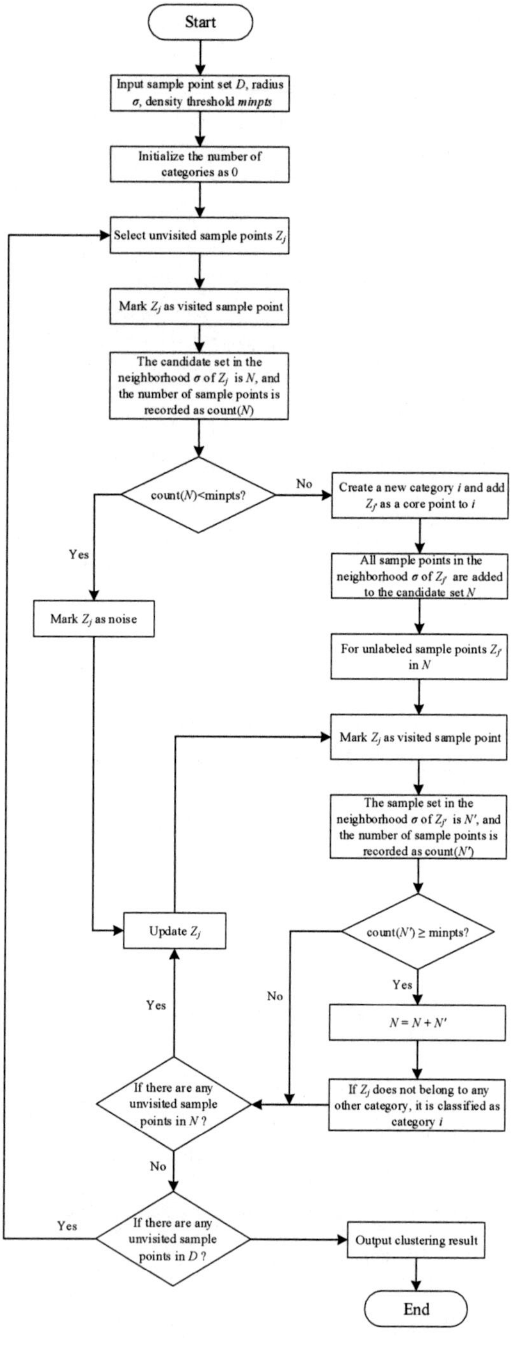

Fig. 5.2 The process of the DBSCAN clustering algorithm

$$
\begin{aligned}
J = & \sum_{i=1}^{I} \sum_{j=1}^{k} \mu_{ij}^{m} \left\| z_j - i_c \right\|^2 \\
& + \lambda_1 \left(\sum_{i=1}^{I} \mu_{ij} - 1 \right) + \cdots \\
& + \lambda_j \left(\sum_{i=1}^{I} \mu_{ij} - 1 \right) + \cdots \\
& + \lambda_k \left(\sum_{i=1}^{I} \mu_{ik} - 1 \right)
\end{aligned}
\tag{5.5}
$$

To find the extreme value of J, the partial derivatives of J to μ_{ij} and i_c equal to 0, there are

$$
\begin{cases}
\frac{\partial J}{\partial \mu_{ij}} = m \left\| z_j - i_c \right\|^2 \mu_{ij}^{m-1} + \lambda_j = 0 \\
\frac{\partial J}{\partial i_c} = \sum_{j=1}^{k} \left[-2\mu_{ij}^{m} (z_j - i_c) \right] = 0
\end{cases}
\tag{5.6}
$$

Then solve to get

$$
\begin{cases}
\mu_{ij} = \dfrac{1}{\sum_{k=1}^{I} \left(\frac{\left\| z_j - i_c \right\|^2}{\left\| z_j - i_k \right\|^2} \right)^{\frac{2}{m-1}}} \\
i_c = \dfrac{\sum_{j=1}^{k} \mu_{ij}^{m}}{\sum_{j=1}^{k} \mu_{ij}^{m}} z_j
\end{cases}
\tag{5.7}
$$

The process of the FCM clustering algorithm is as follows. It is necessary to set the number of clusters i, the weighting index m, and the termination error η to control the end of this loop. Based on the limited conditions of formula (5.4), Randomly divide μ_{ij}, and set the number of iterations and cluster centers.

(1) Calculate the cluster center i_c by formula (5.7).
(2) Update the membership degree μ'_{ij} according to the result of step (1) and formula (5.7).
(3) Execute cyclically Step (1) and Step (2) until the objective function J tends to be stable, and the stable condition is as follows

$$
\varepsilon = \sum_{i=1}^{I} \sum_{j=1}^{k} \left(\mu'_{ij} - \mu_{ij} \right) < \eta
\tag{5.8}
$$

The process of the FCM clustering algorithm is shown in Fig. 5.3.

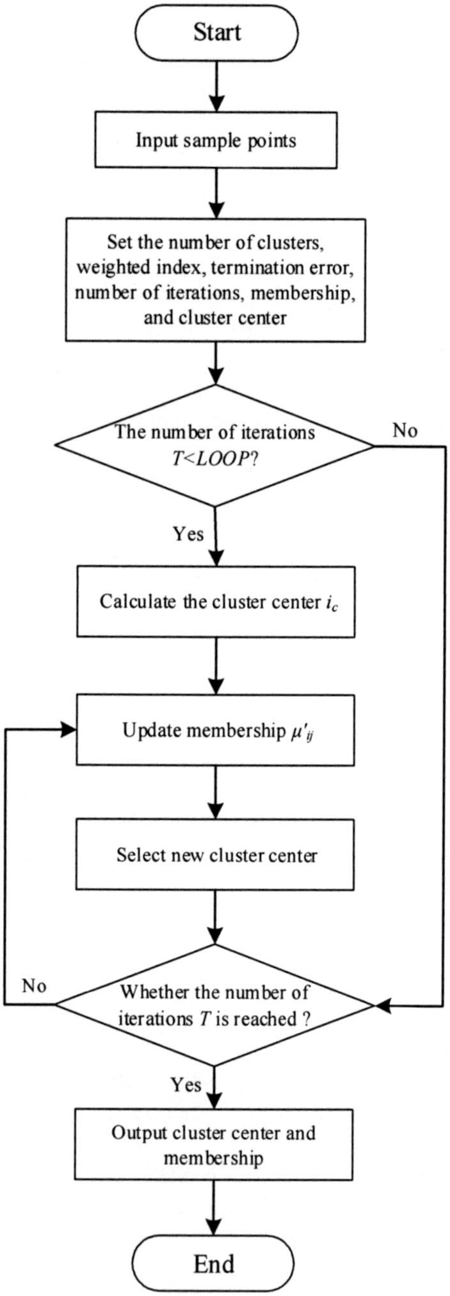

Fig. 5.3 The process of the FCM clustering algorithm

The characteristic of the FCM clustering algorithm is to determine the clustering result by judging the membership degree of the sample points. If a sample point has the highest membership degree in a certain category, the sample point belongs to this category. The FCM clustering algorithm needs to preset some parameters. The selection of these parameters is very important. The selection of parameters is closely related to the quality of the clustering results. When solving the clustering problem of sample points with a large number of features, the FCM clustering algorithm requires a large amount of calculation, which causes poor real-time performance. In the traffic scene analyzed in this chapter, the sample points of adjacent vehicles are relatively close, and the FCM clustering algorithm cannot distinguish the adjacent vehicles in the traffic scene well.

5.1.3 K-Means Clustering Algorithm

K-Means is a clustering algorithm with a simple theory. It utilizes the idea of iterative calculation, randomly selects a sample point from all sample points as the initial clustering center, and calculates the distance between each sample point and the clustering center, the distance is the basis for judging the category of the sample point. The sample point belongs to the category whose cluster center is closest to it. Then calculate the clustering center of the newly divided category, and repeat the above process, until the cluster center no longer changes. The important parameters in the K-Means clustering algorithm include the preset value k, the distance from each sample point to the cluster center, the cluster center of the newly divided category, etc.

In the K-Means clustering algorithm, a large number of "distances" need to be calculated. Suppose the sample point is $z_j(j = 1, 2, \cdots, m_k)$, according to the theory of the K-Means clustering algorithm, before calculating the distance, k sample points need to be randomly selected as the cluster center, and the number of cluster centers is preset by the user, here the value of k is set as the target number in the actual traffic scene.

For the sake of calculating the distance from each sample point to the cluster center, the cluster center is set as $z_i(i = 1, 2, \cdots, k)$, based on Euclidean distance, the specific calculation formula is as follows:

$$\mathrm{dist}\left(z_i, z_j\right) = \sqrt{\sum_{i=1}^{k}\sum_{j=1}^{m_k}\left(z_i - z_j\right)^2} \tag{5.9}$$

The distance from each sample point to the cluster center is the only classification basis of the K-Means clustering algorithm. Here, the distance is defined as the length of the line segment between each sample point and the cluster center in the plane rectangular coordinate system. The concept of distance can be extended to similarity. The closer the distance is, the higher the similarity is. Through the formula (5.9),

k distances can be calculated for each sample point. Then select the cluster center with the shortest distance from one sample point, and divide this sample point into the category of the cluster center, thus, the first sample point classification process is completed. By performing the above operations for each sample point, it can be seen that the preliminary classification process will end after m_k cycles.

Next, calculate the mean value of each category of sample points, and take the mean value as the new clustering center. The algorithm is an iterative process, so the condition for the end of the iteration needs to be set. The condition can be that the cluster center does not change, or the number of iterations reaches the upper limit.

The process of the K-Means clustering algorithm is shown in Fig. 5.4.

The K-Means clustering algorithm is easy to understand and implement. It has high efficiency when dealing with sample point sets with a large amount of data. When the clustering is approximately Gaussian distribution, the effect is better. However, the algorithm needs to preset the number of clusters. The closer the parameter is to the real value, the better the clustering result is. In practice, it is found that the algorithm is very sensitive to noise and outliers. When the sample points of adjacent vehicles are close and mixed with noise, the K-Means clustering algorithm cannot distinguish adjacent vehicles well.

5.2 Spindle-based Density Peaks Fuzzy Clustering Algorithm

In 2014, Rodriguez and Laio proposed the Clustering by Fast Search and Find of Density Peaks (CFSFDP) algorithm. The algorithm considers that, compared with adjacent sample points, the sample points serving as cluster centers have higher density, and each cluster center is far from the sample points with high density. Based on the idea of this algorithm and appropriate innovation, the Spindle-based Density Peaks Fuzzy Clustering (SDPFC) algorithm is proposed in this section. The algorithm corrects the results through secondary clustering. Firstly, the cluster center and cluster category are obtained through the initial Clustering of Density Peak algorithm, and then the clustering results are modified by the Fuzzy Clustering algorithm based on spindle update to obtain the clustering results.

5.2.1 Initial Clustering Algorithm Based on Density Peak

Suppose the sample point set to be clustered is $S = \{z_i\}_{i=1}^{n}$, and $I_S = \{1, 2, \cdots, n\}$ is the corresponding index set. The Euclidean distance between sample points is d_{ij}, the number of d_{ij} is $0.5n(n-1)$, and d_{ij} is

$$d_{ij} = \mathrm{dist}\big(z_i, z_j\big) \tag{5.10}$$

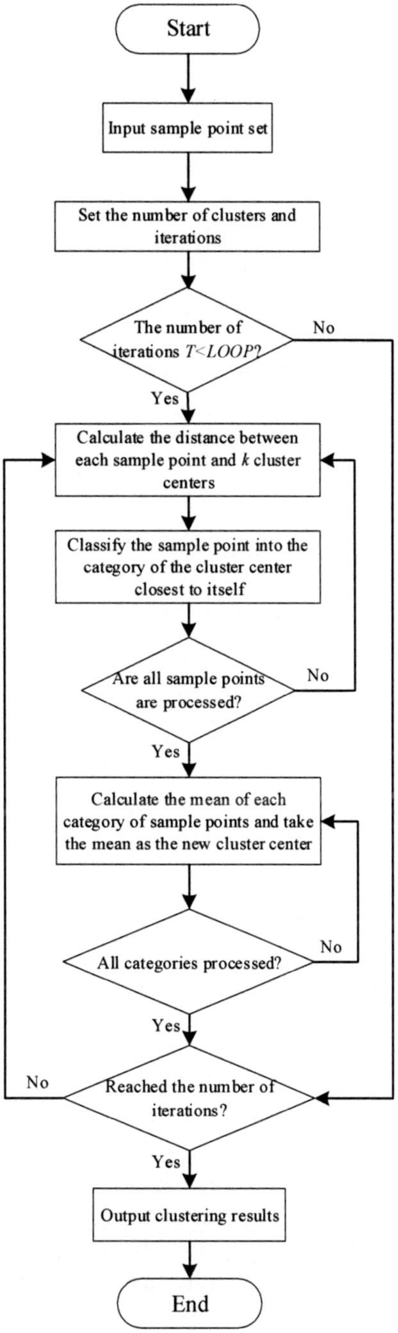

Fig. 5.4 The process of the K-Means clustering algorithm

Arrange d_{ij} in ascending order, define d_t as the r-th d_{ij}, set the percentage parameter p, then r is

$$r = \text{round}\left[\frac{n(n-1)}{2}p\right] \tag{5.11}$$

In the formula, the round() function rounds a number to the nearest integer. In the experimental environment of this section, the sample points are relatively close, Therefore p is chosen to be 2%, and users can modify it according to their own experimental environment. The larger p is chosen, the more cluster categories are filtered.

The density ρ_i of the sample point z_i is

$$\rho_i = \sum_{j \in I_S^i} \chi\left(d_{ij} - d_c\right) \tag{5.12}$$

In the formula, d_c is the truncation distance, which is determined according to the actual scene. When $x \leq 0$, $\chi(x) = 1$; When $x > 0$, $\chi(x) = 0$.

For each sample point z_i, find all sample points z_j, which are denser than the sample point z_i, and select the smallest point among d_{ij}, denoted as δ_i. If there is no sample point z_j denser than z_i, then select the largest point among d_{ij}, denoted as δ_i. Whether the sample point is a cluster center can be determined according to the characteristics of ρ_i and δ_i, δ_i is

$$\delta_i = \begin{cases} \min\limits_{j \in I_S^i}\left\{d_{ij}\right\}, I_S^i \neq \emptyset \\ \max\limits_{j \in I_S^i}\left\{d_{ij}\right\}, I_S^i = \emptyset \end{cases} \tag{5.13}$$

The expression for the indicator set is as follows:

$$I_S^i = \{k \in I_S : \rho_k > \rho_i\} \tag{5.14}$$

Suppose the density threshold is ρ_0, and the distance threshold is δ_0. For the sample point z_i if $\rho_i > \rho_0$ and $\delta_i > \delta_0$, then z_i is considered to be the cluster center of a certain category.

The decision diagram is shown in Fig. 5.5. It can be seen from Fig. 5.5 that the closer the sample points are to the upper right corner, the larger both ρ_i and δ_i are, and these sample points are more likely to be cluster centers. With a decision diagram in place, density thresholds and distance thresholds can be easily defined to determine which points can become cluster centers.

Define $\left\{m_j\right\}_{j=1}^{n_c}$ as the number of sample points corresponding to each cluster center, z_{m_j} as the j-th cluster center. Define $\{c_i\}_{i=1}^{n}$ as the classification attribute mark of the sample point. c_i indicates that the i_{th} sample point belongs to the c_i-th category, which is satisfied

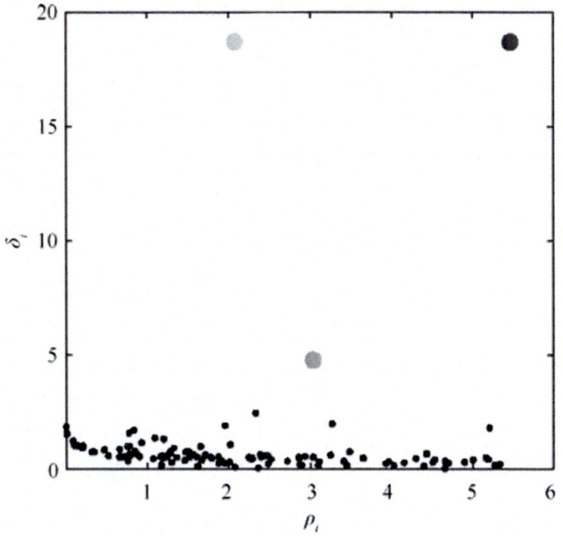

Fig. 5.5 The decision diagram

$$c_i = \begin{cases} k, & z_i \text{ is the } k\text{ - th cluster center} \\ -1, & z_i \text{ is not the } k\text{ - th cluster center} \end{cases} \quad (5.15)$$

Define $\{n_i\}_{i=1}^{n}$ as the number of the nearest sample point among all sample points in S, and whose local density is greater than z_i. It is donated specifically as follows:

$$n_i = \begin{cases} \arg\left(\min_{j \in I_S^i}\{d_{ij}\}\right), & I_S^i \neq \emptyset \\ 0, & I_S^i = \emptyset \end{cases} \quad (5.16)$$

The sample points can be processed by formula (5.16). The sample point with the highest density is closest to which cluster center, and it is classified into this category. This way of processing one by one is much faster than loop iteration.

$\{h_i\}_{i=1}^{n}$ indicates that one sample point is a core sample point (Cluster Core) or an edge sample point (Cluster Halo). The average local density of the core sample points is large, corresponding to the core part of the category, the average local density of the edge sample points is small, corresponding to the edge part of the category, the expression of h_i is

$$h_i = \begin{cases} 0, & z_i \in \text{Cluster Core} \\ 1, & z_i \in \text{Cluster Halo} \end{cases} \quad (5.17)$$

n_c is the number of clusters, if $n_c > 1$, each class will produce a value $\left\{\rho_i^b\right\}_{i=1}^{n_c}$, which is an upper bound on the average local density. For a fixed category, once its boundary area is determined, then the sample points that constitute this area belong to this category, but within the range of distance not exceeding d_c, there are sample points belonging to other categories. Utilizing the boundary area, the average local density can be calculated, and the core sample points and edge sample points can be distinguished.

The average local density is

$$\overline{\rho} = \frac{\rho_i + \rho_j}{2} \tag{5.18}$$

The upper bound of the average local density is

$$\begin{cases} \rho_{c_i}^b = \overline{\rho}, \ \overline{\rho} > \rho_{c_i}^b \\ \rho_{c_j}^b = \overline{\rho}, \ \overline{\rho} > \rho_{c_j}^b \end{cases} \tag{5.19}$$

And h_i is

$$\begin{cases} h_i = 1, \ \rho_i < \rho_{c_i}^b \\ h_i = 0, \ \rho_i \geq \rho_{c_i}^b \end{cases} \tag{5.20}$$

The process of the initial clustering algorithm based on density peaks is as follows.

(1) Initialize and preprocess the data, define truncation distance, calculate d_{ij}, ρ_i and δ_i.
(2) Determine the cluster center, initialize the sample point classification attribute mark, and obtain the cluster center as well as its category number.
(3) Classify the non-cluster center sample points.
(4) If $n > 1$, then divide the sample points in each category into core sample points and edge sample points.

Taking the real traffic scene as an example, the initial clustering results are shown in Fig. 5.6. The horizontal and vertical axes in Fig. 5.6 represent the distance, and the points in the figure represent the sample points. The figure shows the distance between sample points, not the position of actual sample points. The distance between sample points is real, but their coordinates are not real positions. The figure is only used to clearly show the cluster center and cluster category.

5.2.2 *Fuzzy Clustering Algorithm Based on Spindle Update*

To obtain the final clustering results, the initial clustering results need to be modified in this section. The results used in this section include the coordinates of each sample point, the location of each clustering center, and the number of clustering categories. i represents the i-th sample category ($i = 1, 2, \cdots, I$), and $z_0^{(i)}$ represents the cluster center of the i-th sample category.

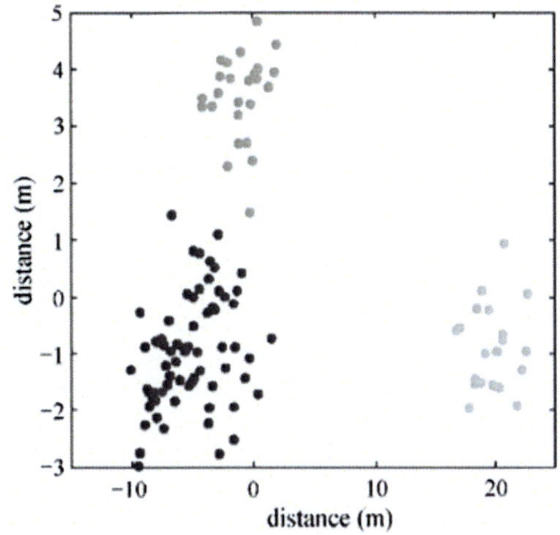

Fig. 5.6 The initial clustering results

The millimeter-wave surveillance radar is adopted in the simulation scene, which is suspended at the bayonet above the lane. The position of the radar and lane is fixed. Under normal situations, it is illegal for a vehicle to drive between lane lines, step on the line for a long time, or cross the line illegally. Therefore, based on the results obtained in Sect. 5.1, take the straight line passing through the clustering center and parallel to the lane centerline as the spindle, to modify the initial clustering results. Make a special calibration vehicle drive on the road, collect the position information of the centerline of a certain lane, and record the slope of the straight line as a_0.

Define the initial spindle as a straight line over the cluster center and parallel to the lane centerline, the i-th spindle $L_c^{(i)}$ is

$$y = a^{(i)}x + b^{(i)} \tag{5.21}$$

Suppose the initial spindle passes through each cluster center $z_0^{(i)} = \left(x_0^{(i)}, y_0^{(i)}\right)$, and its slope is a_0, there is

$$y = a_0^{(i)}x + b_0^{(i)} \tag{5.22}$$

Define the coordinates of the sample point in the plane Cartesian coordinate system as

$$z_j = \left(x_j, y_j\right) \tag{5.23}$$

$j = 1, 2, \cdots, k$ represents the j-th sample point except for the cluster center, and k represents the number of all sample points except for the cluster center.

Equation (5.24) represents the distance between the j-th sample point and the i-th spindle, its value is $D_j^{(i)}$.

$$z_j \sim L_c^{(i)} \tag{5.24}$$

Define the objective function T as

$$T = \sum_{i=1}^{I} \sum_{j=1}^{k} B_{ij}^m D_j^{(i)^2} \tag{5.25}$$

B is the fuzzy matrix. The fuzzy matrix is the probability matrix that a sample point belongs to a certain category. m represents the degree to which a sample point belongs to a certain category, i.e. weight. It can be determined by the user. The constraint condition of B is

$$\sum_{i=1}^{I} B_{ij} = 1 \tag{5.26}$$

To obtain the optimal solution, it is necessary to make T a minimum value under the constraints. Lagrangian expansion of Eq. (5.25), there is

$$\begin{aligned} T = & \sum_{i=1}^{I} \sum_{j=1}^{k} B_{ij}^m D_j^{(i)^2} + \lambda_1 \left(\sum_{i=1}^{I} B_{ij} - 1 \right) \\ & + \cdots + \lambda_j \left(\sum_{i=1}^{I} B_{ij} - 1 \right) + \cdots + \lambda_k \left(\sum_{i=1}^{I} B_{ik} - 1 \right) \end{aligned} \tag{5.27}$$

Find the partial derivative of T over B_{ij} and get

$$\frac{\partial T}{\partial B_{ij}} = m D_j^{(i)^2} B_{ij}^{m-1} + \lambda_j = 0 \tag{5.28}$$

Obtain the following by calculation

$$\begin{aligned} B_{ij}^{m-1} = & \frac{-\lambda_j}{m D_j^{(i)^2}} \\ \Rightarrow B_{ij} = & \left(\frac{-\lambda_j}{m} \right)^{\frac{1}{m-1}} \frac{1}{\left[D_j^{(i)^2} \right]^{\frac{2}{m-1}}} \end{aligned} \tag{5.29}$$

Substitute B_{ij} into Eq. (5.26), there is

$$\left(\frac{-\lambda_j}{m} \right)^{\frac{1}{m-1}} \sum_{i=1}^{I} \frac{1}{\left[D_j^{(i)^2} \right]^{\frac{2}{m-1}}} = 1 \tag{5.30}$$

Simplify and get

$$\left(\frac{-\lambda_j}{m}\right)^{\frac{1}{m-1}} = \frac{1}{\sum_{i=1}^{I} \frac{1}{\left[D_j^{(i)2}\right]^{\frac{2}{m-1}}}} = \frac{1}{\sum_{k=1}^{I} \frac{1}{\left[D_j^{(k)2}\right]^{\frac{2}{m-1}}}} \tag{5.31}$$

Then substitute Eq. (5.31) into Eq. (5.29), there is

$$\begin{aligned} B_{ij} &= \frac{1}{\sum_{k=1}^{I} \frac{1}{\left[D_j^{(k)2}\right]^{\frac{2}{m-1}}}} \frac{1}{\left[D_j^{(i)2}\right]^{\frac{2}{m-1}}} \\ &= \frac{1}{\sum_{k=1}^{I} \left[\frac{D_j^{(i)2}}{D_j^{(k)2}}\right]^{\frac{2}{m-1}}} \end{aligned} \tag{5.32}$$

It can be seen from Eq. (5.32) that to obtain B_{ij}, the distance from the sample point i to the cluster center j should be calculated first. Secondly, calculate the sum of the distances from the sample point i to all the cluster centers, then find the ratio of the former to the latter. The higher the value, the higher the membership degree. Equation (5.32) is the updated formula of B_{ij}, and the spindle update process is as follows.

According to the previous definition, the spindle expression is $y = a^{(i)}x + b^{(i)}$. The sample points z_j are $(x_1, y_1), (x_2, y_2), \cdots, (x_{m_k}, y_{m_k})$.

$D_j^{(i)}$ is the distance from the sample point to the i-th spindle, i.e. error value. There is

$$D_j^{(i)} = \begin{cases} \frac{\left|a^{(i)}x_j - y_j + b^{(i)}\right|}{\sqrt{a^{(i)2}+1}}, & y_0^{(i)} - \frac{d_c}{\sin\{\arctan[a^{(i)}]\}} \le y_j \le y_0^{(i)} + \frac{d_c}{\sin\{\arctan[a^{(i)}]\}} \\ \sqrt{\left[x_j^{(i)2} - x_j^2\right] + \left[y_j^{(i)2} - y_j^2\right]}, & \text{others} \end{cases} \tag{5.33}$$

If the ordinate of the sample point position is within the distance d_c above and below the ordinate of a certain cluster center, calculate the distance from the sample point to a certain spindle, otherwise calculate the distance from the sample point to a certain cluster center. d_c is generally taken as the length of half the car, and the specific value needs to be tested in practical applications. When $D_j^{(i)}$ is the minimum value, the sample point can be classified as the i-th category.

Define δ as the sum of squares of errors, and its expression is

$$\delta = \sum_{j=1}^{m} D_j = \sum_{j=1}^{m} \left(y_j - ax_j - b\right)^2 \tag{5.34}$$

In which, m represents the number of sample points in a category except for the cluster center. Calculate the partial derivative of δ over a and b respectively, there are

$$\begin{cases} \frac{\partial \delta}{\partial a} = -2\left(\sum_{j=1}^{m} x_j y_j - b\sum_{j=1}^{m} x_j - a\sum_{j=1}^{m} x_j^2\right) \\ \frac{\partial \delta}{\partial b} = -2\left(\sum_{j=1}^{m} y_j - \sum_{j=1}^{m} b - a\sum_{j=1}^{m} x_j\right) \end{cases}$$
$$\Rightarrow \begin{cases} \frac{\partial \delta}{\partial a} = -2\left(\sum_{j=1}^{m} x_j y_j - bm\overline{x} - a\sum_{j=1}^{m} x_j^2\right) \\ \frac{\partial \delta}{\partial b} = -2(m\overline{y} - bm - am\overline{x}) \end{cases} \tag{5.35}$$

Formula (5.35) equal to 0, it can be obtained

$$\begin{cases} a = \frac{\sum_{j=1}^{m}(x_j - \overline{x})(y_j - \overline{y})}{\sum_{j=1}^{m}(x_j - \overline{x})^2} \\ b = \overline{y} - a\overline{x} \end{cases} \tag{5.36}$$

The process of the fuzzy clustering algorithm based on spindle update is as follows.

(1) Input parameters: sample set $z_j^{(i)}$, initial spindle slope a_0, initial clustering category i, the spindle of each category $L_c^{(i)}$.
(2) Update matrix B.
(3) Update the spindle of each category of sample points.
(4) Repeat steps (2) and (3) until the variable $V < \eta$.

The expression of V is

$$V = \sum_{i=1}^{I} \sum_{j=1}^{m_k} \left(B'_{ij} - B_{ij}\right) \tag{5.37}$$

Figure 5.7 shows the process of the spindle-based density peak fuzzy clustering algorithm. Taking the experimental scene in Sect. 5.1 as an example, the clustering results are shown in Fig. 5.8. The sample points in the figure are collected from a real highway traffic scene, and the coordinates of the sample points correspond to the coordinates in the real scene.

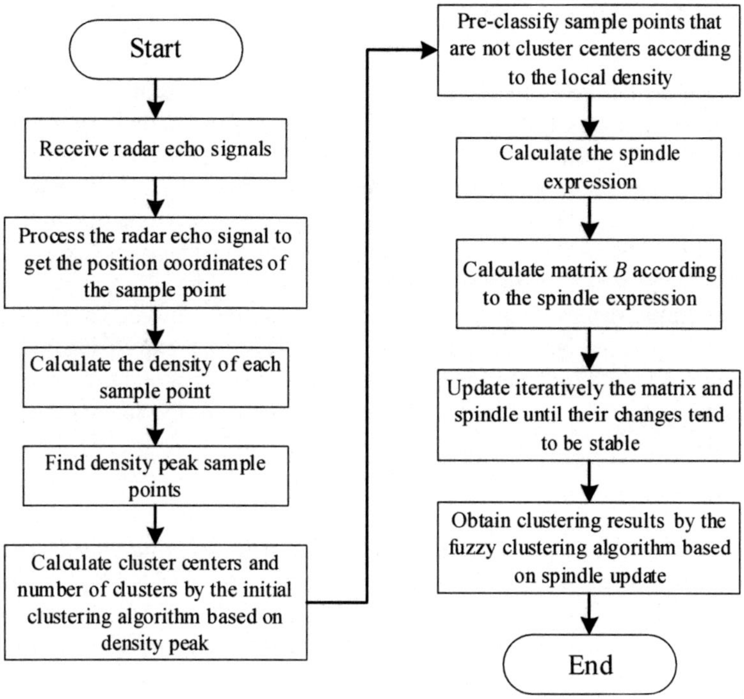

Fig. 5.7 The process of the spindle-based density peak fuzzy clustering algorithm

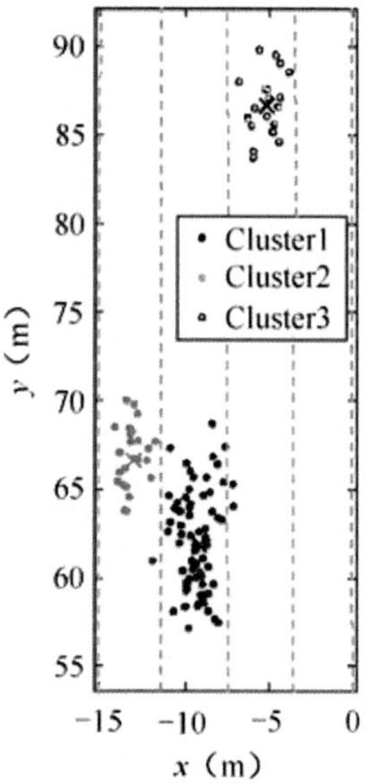

Fig. 5.8 The clustering results

5.2.3 *Experimental Design*

The experimental scene is a four-lane highway, the radar is installed on the top of the bayonet side, and the position of the radar is shown in Fig. 5.9. In the highway traffic scene, the vehicle runs at a fast speed and the distance between the front and rear vehicles is large. During driving, the vehicles will approach and block each other. Existing clustering algorithms cannot accurately distinguish adjacent and mutually occluded objects. In this section, the spindle-based density peaks fuzzy clustering algorithm is proposed, which can better solve the problem of adjacent vehicles in highway traffic scenes. From the positional relationship between the radar and the road in this experimental scene, it can be get that the spindle expression is $x = \text{center}(x)$, where center (x) represents the abscissa of a cluster center. In the experiments, the spindle of the experiment remains unchanged while the main scene remains unchanged.

1. Scenario 1

On a four-lane expressway, there are three vehicles within close driving distance. Two of the large vehicles are laterally close to each other and block each other, the other vehicle is slightly farther away. Scenario 1 is shown in Fig. 5.10. The radar returns a total of 117 valid sample points, and the sample point distribution of Scenario 1 is shown in Fig. 5.11.

In Scenario 1, two large vehicles block each other, resulting in uneven distribution of sample points. The DBSCAN clustering algorithm classifies according to the density and obtains many categories; the FCM clustering algorithm iteratively updates the fuzzy matrix and the cluster center, and cannot distinguish the adjacent objects well; the K-Means clustering algorithm also cannot distinguish the adjacent objects well. The results obtained by the SDPFC clustering algorithm correspond to the real scene, and the effect is better. Figure 5.12 shows the clustering results of each algorithm in Scenario 1.

Fig. 5.9 The position of the radar

Fig. 5.10 Scenario 1

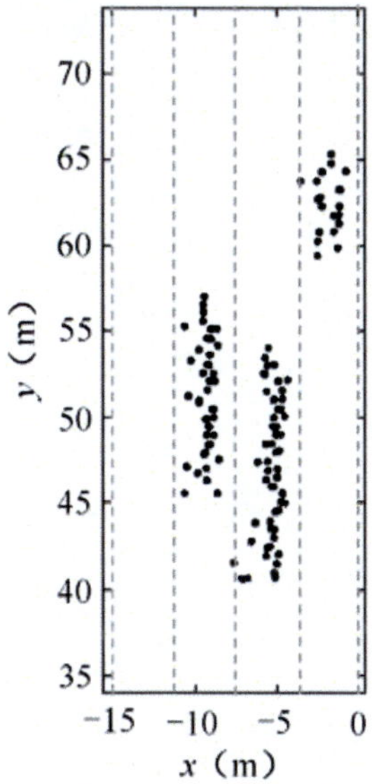

Fig. 5.11 The sample point distribution of Scenario 1

2. Scenario 2

On a four-lane expressway, there are three vehicles within close driving distance. Two of the vehicles are very close laterally and the large vehicle blocks the small vehicle, the other vehicle is slightly farther away from the two vehicles. Scenario 2 is shown in Fig. 5.13. The radar returns a total of 115 valid sample points, and the sample point distribution of Scenario 2 is shown in Fig. 5.14.

The clustering results of each algorithm in Scenario 2 are shown in Fig. 5.15.

In Scenario 2, a large car blocks a small car, resulting in a small number of sample points for small cars. Compared with the DBSCAN clustering algorithm, FCM clustering algorithm, and K-Means clustering algorithm, the results obtained by the SDPFC clustering algorithm are better.

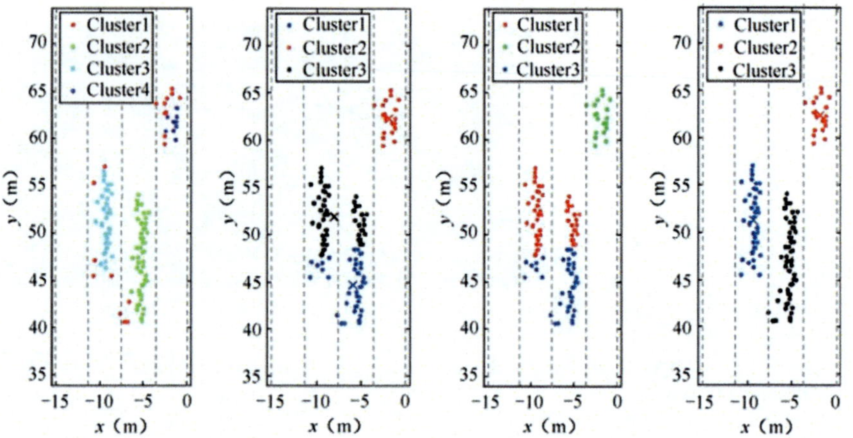

(a) DBSCAN (b) FCM (c) K-Means (d) SDPFC

Fig. 5.12 The clustering results of each algorithm in Scenario 1

Fig. 5.13 Scenario 2

Fig. 5.14 The sample point distribution of Scenario 2

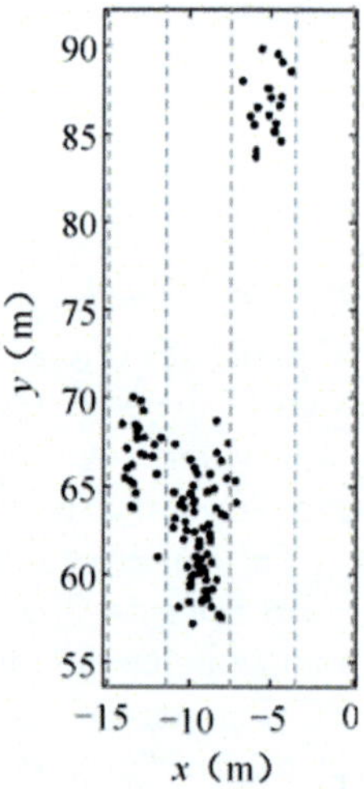

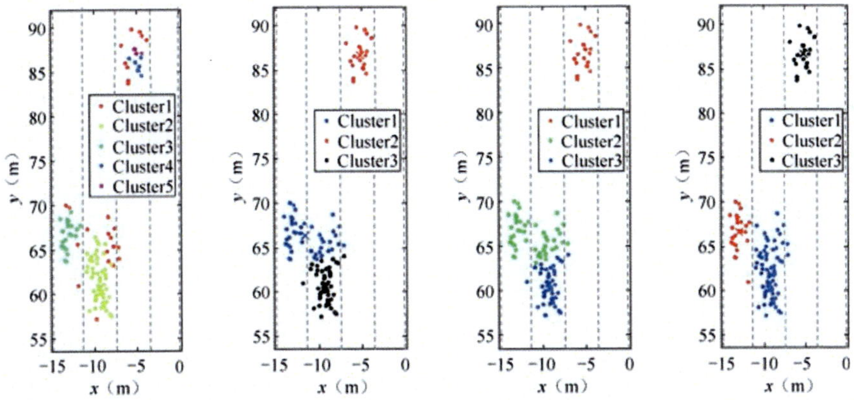

(a) DBSCAN (b) FCM (c) K-Means (d) SDPFC

Fig. 5.15 The clustering results of each algorithm in Scenario 2

3. Scenario 3

On a four-lane expressway, five vehicles travel close together, and multiple vehicles block each other. Scenario 3 is shown in Fig. 5.16. The radar returns a total of 202 valid sample points. The sample point distribution of Scenario 3 is shown in Fig. 5.17.

The clustering results of each algorithm in Scenario 3 are shown in Fig. 5.18.

In Scenario 3, the number of adjacent samples is large, and many vehicles are blocking each other, resulting in uneven distribution of sample points and a little number of sample points for small cars. Compared with the DBSCAN clustering algorithm, FCM clustering algorithm, and K-Means clustering algorithm, the results obtained by the SDPFC clustering algorithm are better.

To visually compare the accuracy of each algorithm, the clustering accuracy is defined as

$$\text{Accuracy} = \frac{K}{N} \tag{5.38}$$

Fig. 5.16 Scenario 3

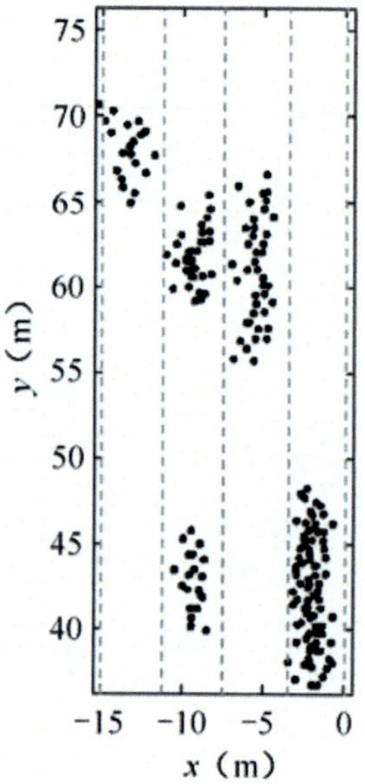

Fig. 5.17 The sample point distribution of Scenario 3

where K is the number of correctly classified sample points and N is the number of sample points involved in the classification. The comparison of the clustering accuracy of each algorithm is shown in Fig. 5.19.

As can be seen from Fig. 5.19, in the three actual scenarios in this section, the SDPFC clustering algorithm has the highest clustering accuracy, with an average clustering accuracy of more than 0.95.

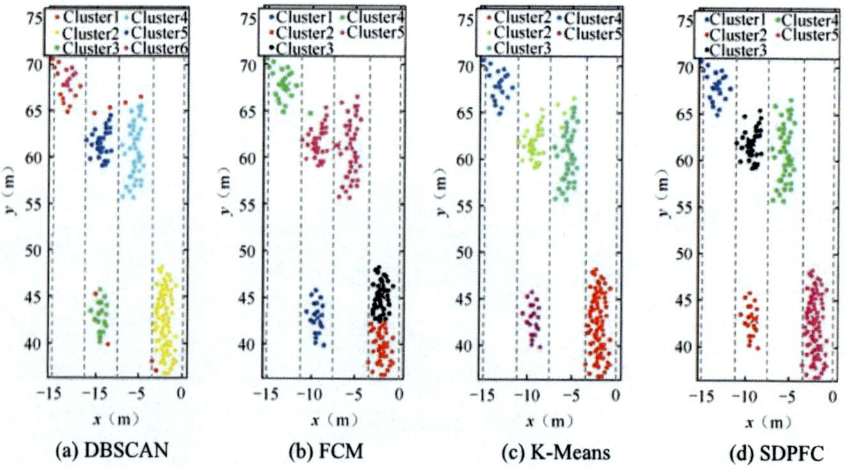

Fig. 5.18 The clustering results of each algorithm in Scenario 3

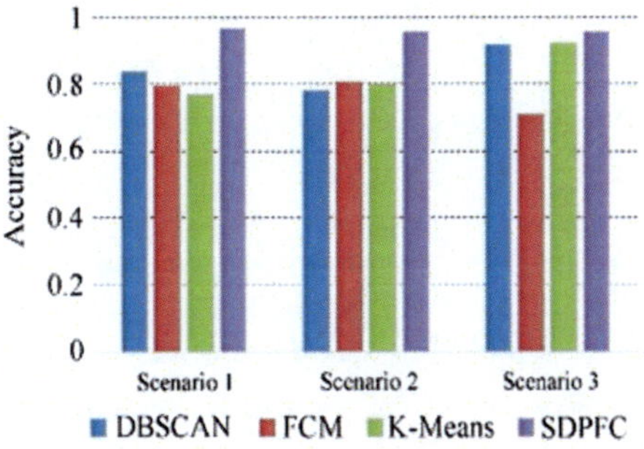

Fig. 5.19 The comparison of the clustering accuracy of each algorithm

To better present the clustering results of the SDPFC clustering algorithm, in addition to the above three classic scenarios, this section also selects four scenarios with adjacent vehicles, and the real situations of the four scenarios are shown in Fig. 5.20. Figure 5.21 shows the clustering result of the SDPFC clustering algorithm.

To visually compare the real-time performance of each algorithm, four algorithms are tested, and all sample points are selected as the input data of the experiment in each scene within about 0.2 s. The time consumption of each algorithm is shown in Table 5.1.

It can be seen from Table 5.1 that the DBSCAN clustering algorithm takes the shortest time, the K-Means clustering algorithm ranks second because of its simple calculation, and the FCM clustering algorithm takes the longest. The initial point in the FCM clustering algorithm Uncertainty results in a higher number of iterative updates.

(a) Scenario 4

(b) Scenario 5

(c) Scenario 6

(d) Scenario 7

Fig. 5.20 The real situations of the four scenarios

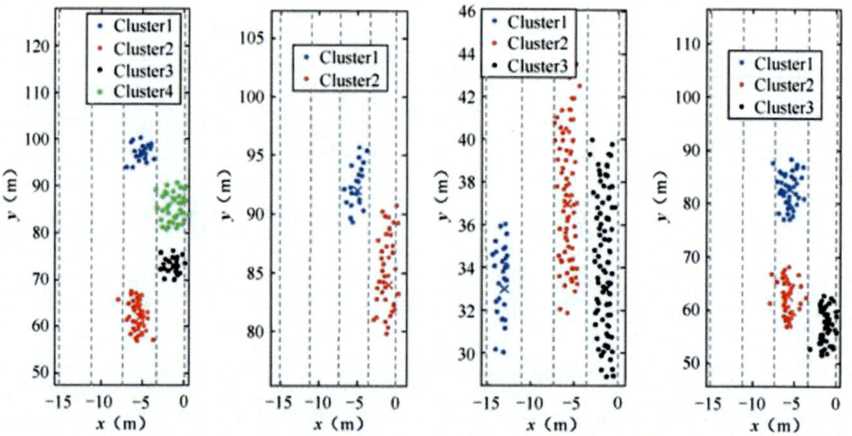

(a) Scenario 4 (b) Scenario 5 (c) Scenario 6 (d) Scenario 7

Fig. 5.21 The clustering result of the SDPFC clustering algorithm

Table 5.1 The time consumption of each algorithm

Number of experiments	Time consumption(s)			
	DBSCAN	FCM	K-Means	SDPFC
1	0.0139	0.0358	0.0270	0.0330
2	0.0128	0.0318	0.0296	0.0298
3	0.0141	0.0337	0.0262	0.0322
4	0.0138	0.0325	0.0265	0.0351
5	0.0129	0.0358	0.0254	0.0325
6	0.0132	0.0334	0.0259	0.0329
7	0.0135	0.0315	0.0249	0.0332
8	0.0139	0.0324	0.0266	0.0336
9	0.0134	0.0349	0.0271	0.0332
10	0.0136	0.0338	0.0256	0.0319
Average	0.0135	0.0336	0.0266	0.0327

After comparing the clustering accuracy and running speed of each algorithm in a specific scenario, the stability comparison of each algorithm is shown in Fig. 5.22. As can be seen from Fig. 5.22, the SDPFC clustering algorithm has strong adaptability and high clustering accuracy in specific scenarios; the K-Means clustering algorithm is more adaptable, but the clustering accuracy is not high; The adaptability of the DBSCAN clustering algorithm and FCM clustering algorithm is relatively weak, but the clustering accuracy of DBSCAN is obviously higher than FCM.

To better show the clustering results of the SDPFC clustering algorithm, Additional experiments are presented. Statistics on the accuracy of clustering adjacent

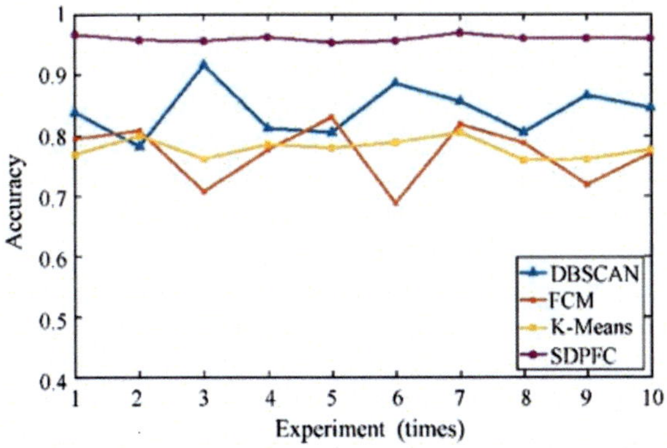

Fig. 5.22 The stability comparison of each algorithm

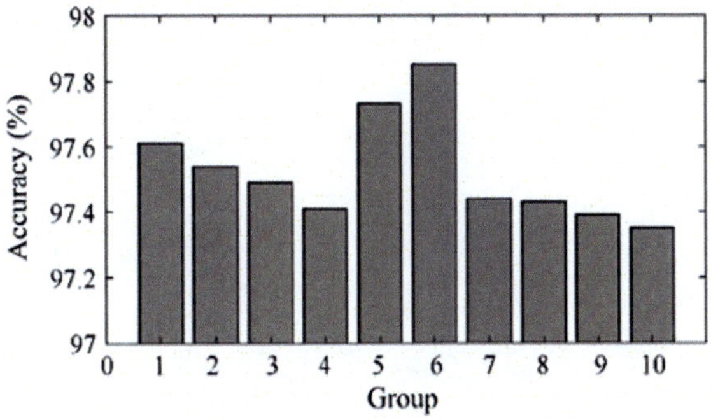

Fig. 5.23 The clustering accuracy of the SDPFC algorithm in 1000 scenes

vehicles in 1000 scenes. Vehicle sample points collected every 40 ms are defined as a scene. In each scene, 2–5 vehicles are passing by, and process the radar echo signal to obtain 100–300 sample points. The clustering accuracy is calculated by comparing each scene with real images and clustering results. Figure 5.23 shows the clustering accuracy of the SDPFC algorithm in 1000 scenes. According to the statistics, 3627 passing vehicles were processed in 1000 scenes. Take 100 scenes as a group for discrimination, and 3537 vehicles obtained the correct clustering results. The accuracy of clustering is about 97.52%.

In the highway traffic scene, the SDPFC clustering algorithm can solve the problem of inaccurate clustering results of adjacent vehicles. The algorithm takes less time, has high precision, and has strong adaptability to scene changes.

5.3 Summary

A Spindle-based Density Peaks Fuzzy Clustering (SDPFC) algorithm is proposed in this chapter. In addition, this chapter discusses the advantages of the clustering algorithm and three classical clustering algorithms: DBSCAN clustering algorithm, FCM clustering algorithm, and K-Means clustering algorithm. For the problem of distinguishing adjacent objects in highway traffic scenes, the classical clustering algorithm is improved, and the SDPFC clustering algorithm is proposed. After analyzing the sample point characteristics of vehicles in the experimental scene, it is found that each sample point cluster has a density peak. Taking advantage of this feature, the initial clustering algorithm based on density peaks is used to obtain the number of clusters and cluster centers, and then the fuzzy clustering algorithm based on spindle update is used to obtain the final clustering results. This chapter compares the clustering results of each algorithm in seven typical scenarios. The results show that the SDPFC clustering algorithm can solve the clustering problem of adjacent vehicles and occluded vehicles in the highway traffic scene, with high accuracy, real-time performance, and robustness.

Chapter 6
Data Association Algorithms

6.1 Improved Particle Filtering Algorithm

6.1.1 Track Association

Track association means that when the radar tracks the target in real time, the target detected in the radar echo signal at the current moment is matched with the target track that already exists at the current moment. The multi-target track association algorithm usually adopts the Nearest Neighbor (NN) algorithm, the Joint Probability Data Association (JPDA) algorithm, and the Multiple Hypothesis Tracking (MHT) algorithm. The algorithm comparison is shown in Table 6.1 shown.

Radar is mainly used to monitor the road environment, so various types of vehicles are usually tracked objects. According to the physical model of the vehicle, echoes from different vehicles can be easily distinguished. We decompose the track association into two steps: one is to select effective targets through the density clustering algorithm and fuse the target points from the same vehicle; the other is to effectively screen the targets based on the NN algorithm. In detail, it is to weight the multi-dimensional information such as the spatial distance, speed, and angle of the target to obtain the matching combination of the candidate target and the tracked target, so as to convert the association problem into an assignment problem, and finally use the Hungarian assignment algorithm to obtain the optimal match. The goal of track association can be achieved through the above two steps.

1. Application of DBSCAN clustering algorithm

In the actual road environment, it is inevitable that the same vehicle has multiple strong reflection centers, resulting in a vehicle being detected as multiple targets. The target detection scene is shown in Fig. 6.1, and the target detection results are shown in Fig. 6.2.

In Fig. 6.1, both the near minivan and the distant truck are detected as two objects, corresponding to the two overlapping detection boxes at 5 m and 45 m in Fig. 6.2, respectively. Note that the two overlapping detection boxes at 20 m represent two

L. Cao et al., *Target Recognition and Tracking for Millimeter Wave Radar in Intelligent Transportation*, https://doi.org/10.1007/978-981-99-1533-0_6

Table 6.1 Algorithm comparison

Algorithm	Advantages	Disadvantages
NN	Small amount of calculation Simple structure	It is prone to false associations when the target density is high additional restrictions need to be added
JPDA	Adapt to dense environment Strong anti-clutter ability	Large amount of calculation poor time performance in practical applications
MHT	Adapt to dense environment Strong anti-clutter ability	Large amount of calculation Excessive reliance on prior information poor time performance in practical applications

Fig. 6.1 Object detection scene

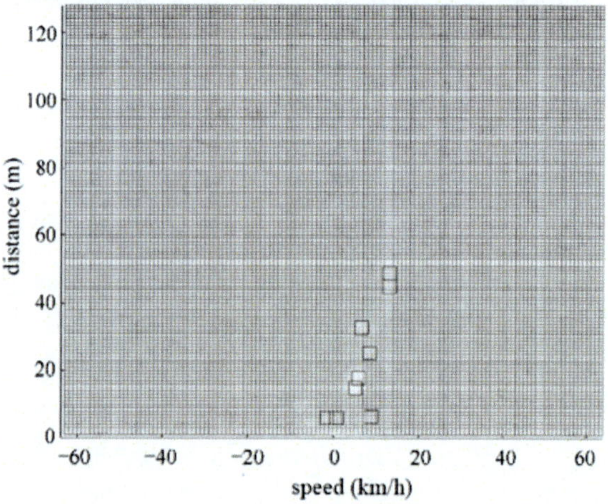

Fig. 6.2 Target detection result

different vehicles. When performing multi-cycle matching, the above situation will reduce the stability and confidence of the track, thereby affecting the tracking effect.

The above problems can be solved by the DBSCAN clustering algorithm, which mainly uses the density characteristics of the samples, takes the samples with low density as the boundary, and classifies the samples with connected densities into one category according to the density characteristics and correlation.

Assume that the data set after target detection is $\boldsymbol{X} = \{\boldsymbol{x}^{(1)}, \boldsymbol{x}^{(2)}, \cdots, \boldsymbol{x}^{(m)}\}$, $\boldsymbol{x}^{(i)} \in R^n$, $\boldsymbol{x}^{(i)}$ is an n-dimensional vector, representing the n-dimensional information of the $i-th$ target, and m is the number of detected targets.

Set the cutoff distance d_c, take any target as the center, and set the cutoff distance d_c as the radius, and set the neighborhood ε of the target. If there are ρ other targets in the neighborhood of this target, the local density of this target is ρ, and the local density of the $i-th$ target is defined as

$$\rho_i = \sum_j \chi(d_{ij} - d_c) \tag{6.1}$$

$$d_{ij} = f(R_i, R_j, V_i, V_j, \theta_i, \theta_j) \tag{6.2}$$

In the formula, d_{ij} represents the distance between $\boldsymbol{x}^{(i)}$ and $\boldsymbol{x}^{(j)}$, d_c represents the cutoff distance, χ represents the relationship between d_{ij} and d_c, and R_i, V_i and θ_i represent the distance, speed, and angle of the $i-th$ target, respectively. The same is true for R_j, V_j and θ_j. f means to calculate multi-dimensional distance.

When applying the DBSCAN clustering algorithm, the following rules need to be followed.

(1) For the sample $\boldsymbol{x}^{(i)}$, if its local density is greater than or equal to the preset sample threshold minpts, the sample $\boldsymbol{x}^{(i)}$ is recorded as the core sample. If its local density is less than the preset sample threshold minpts but greater than 0, the sample $\boldsymbol{x}^{(i)}$ is recorded as a boundary sample. If its local density is equal to the preset sample threshold minpts, the sample $\boldsymbol{x}^{(i)}$ is recorded as a noise sample.
(2) For samples $\boldsymbol{x}^{(i)}$ and $\boldsymbol{x}^{(j)}$, if sample $\boldsymbol{x}^{(j)}$ is included in the neighborhood ε of sample $\boldsymbol{x}^{(i)}$, and $\boldsymbol{x}^{(i)}$ is the core sample, then $\boldsymbol{x}^{(j)}$ is said to be directly connected by the density of $\boldsymbol{x}^{(i)}$.
(3) For a sample sequence $\boldsymbol{x}^{(1)}, \boldsymbol{x}^{(2)}, \cdots, \boldsymbol{x}^{(m)}$, if all the samples in the sample sequence satisfy that $\boldsymbol{x}^{(i+1)}$ is directly connected by the density of $\boldsymbol{x}^{(i)}$, then any $\boldsymbol{x}^{(i)}$ in the sample can be density-reachable from $\boldsymbol{x}^{(1)}$. If $\boldsymbol{x}^{(i+1)}$ is directly connected by the density of $\boldsymbol{x}^{(i)}$, then $\boldsymbol{x}^{(i+1)}$ must be density-reachable from $\boldsymbol{x}^{(i)}$. The inverse of the above proposition does not necessarily hold.
(4) For samples $\boldsymbol{x}^{(i)}$ and $\boldsymbol{x}^{(j)}$, if there is at least one core sample $\boldsymbol{x}^{(k)}$, so that sample $\boldsymbol{x}^{(i)}$ is density-reachable from core sample $\boldsymbol{x}^{(k)}$ and sample $\boldsymbol{x}^{(j)}$ is density-reachable from core sample $\boldsymbol{x}^{(k)}$, then sample $\boldsymbol{x}^{(i)}$ and sample $\boldsymbol{x}^{(j)}$ are density-connected.

(5) For the sample sequence $\boldsymbol{x}^{(1)}, \boldsymbol{x}^{(2)}, \cdots, \boldsymbol{x}^{(m)}$, if the samples in the sample sequence are density-connected, the total set of samples contained in the neighborhood ε of all samples in the sample sequence is called the sample cluster.

We explain the above rules intuitively and clearly with an example. The example contains two sample clusters, in which the white points are core sample points, the black points are non-core sample points, and the gray points are noise points. The sample threshold was set to minpts = 5. Arrows indicate that one of the sample points is directly connected to the other sample points by the density. The diagram of DBSCAN clustering algorithm is shown in Fig. 6.3.

Use the DBSCAN clustering algorithm to cluster the target detection results in Fig. 6.2, and the clustering results are shown in Fig. 6.4.

As can be seen from Fig. 6.4, after clustering using the DBSCAN clustering algorithm, two detection results from one vehicle are merged into one valid target. In practical applications, the clustering algorithm can effectively reduce the impact of

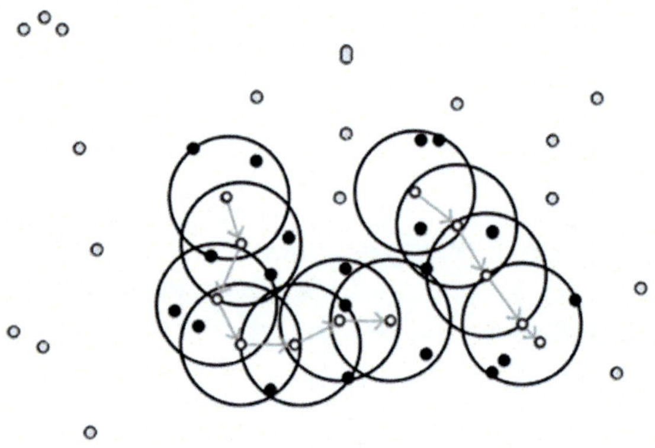

Fig. 6.3 Diagram of DBSCAN clustering algorithm

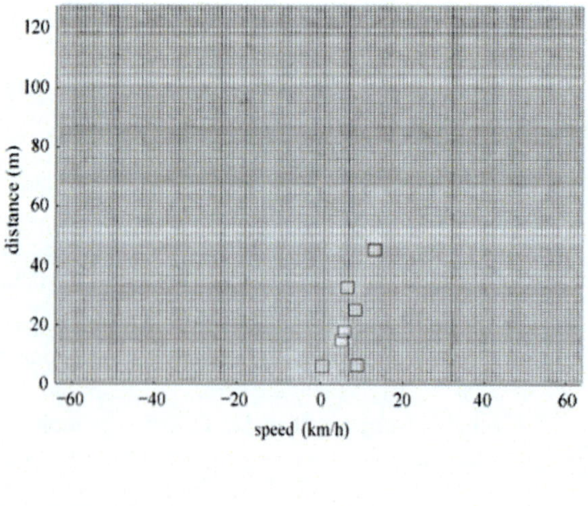

Fig. 6.4 Clustering results

such problems on the stability and reliability of the track, and optimize the tracking effect of the radar.

2. Track association based on Hungarian assignment

After obtaining the valid target, it is necessary to match the valid target with the existing track to provide conditions for filtering. Radars often face complex scenarios with multiple targets. If there are too many detected targets, the calculation amount of the JPDA algorithm and the MHT algorithm will increase seriously, which cannot meet the needs of real-time processing. Therefore, the NN algorithm with additional constraints is used for processing, and the Hungarian assignment algorithm is integrated to achieve the goal of improving the matching efficiency.

For practical scenarios, the number of new targets is usually not equal to the number of existing tracks, that is, $m \neq n$. To match m targets with n tracks, an evaluation matrix needs to be constructed to measure the similarity between targets and tracks. The following is based on the NN algorithm, using the state of the valid target and the state of the existing track to generate a matching covariance matrix.

The state of the valid target includes distance, speed, angle, acceleration and energy, etc. The state of the existing track includes distance, speed, angle, acceleration and energy, etc. The matching covariance C_{ij} is

$$C_{ij} = d(X_i, Z_j) \tag{6.3}$$

he state of the j-th track in the previous period, and the function d represents the multi-dimensional space weighted Euclidean distance of each state (distance, speed, angle, acceleration and energy) of X_i and Z_j. C_{ij} represents the covariance of the i-th effective target and the j-th track in the previous period. The assignment evaluation matrix $\boldsymbol{c}$ is

$$\boldsymbol{C} = \begin{bmatrix} c_{11} & c_{12} & \cdots & c_{1n} & 0 \\ c_{21} & c_{22} & \cdots & c_{2n} & 0 \\ \vdots & \vdots & & \vdots & \vdots \\ c_{m1} & c_{m2} & \cdots & c_{mn} & 0 \end{bmatrix} \tag{6.4}$$

Next, the problem of finding the optimal solution of a matrix is converted into an assignment problem. The assignment problem can be described by the following mathematical model

$$\min z = \sum_{i,j=1}^{n} c_{ij} x_{ij} \tag{6.5}$$

$$\begin{cases} \sum_{i=1}^{n} x_{ij} = 1 \\ \sum_{j=1}^{n} x_{ij} = 1 \\ x_{ij} = 0\text{或}x_{ij} = 1 \end{cases} \tag{6.6}$$

In the formula, $i, j = 1, 2, \ldots, n$, z is the assignment error, and x_{ij} indicates whether to assign or not. The target can be assigned based on the assignment evaluation matrix to obtain the smallest assignment error, and the optimal assignment result can be obtained by using the Hungarian assignment algorithm. On this basis, a threshold is set. If the difference between the target and the track exceeds this threshold, no matching is performed.

The priority of the track of the previous period can be calculated according to the number of periods and the accumulated error of the track of the previous period, and the priority U_k^j is

$$U_k^j = \frac{1}{\sqrt{2\pi}} \exp(T_k^j, E_k^j) \tag{6.7}$$

In the formula, k is the current period, the value is a positive integer, T_k^j is the number of existing periods of the j-th track in the previous period, and E_k^j is the accumulated error of the j-th track in the previous period.

A threshold is set. If U_k^j is higher than this threshold, it is considered that the track has a high degree of confidence, and the optimal value of the corresponding row and column is preferentially selected as the matching result, and the dimension of the assignment evaluation matrix is reduced, and the row and column where the track j is located are removed. Dimension reduction is performed according to the order of priority from high to low, and the Hungarian assignment algorithm is applied to the assignment evaluation matrix after dimension reduction. The Hungarian assignment algorithm flow is shown in Table 6.2.

6.1.2 Moving Target Tracking

After the track association is realized, the effective target state is used as the measurement, and the track state is used as the target estimate for filtering, and the updated target estimate is obtained to realize the target tracking.

Vehicle radar moving target tracking usually adopts Kalman filtering (KF), extended Kalman filtering (EKF), unscented Kalman filtering (UKF), etc. However, in practical applications, since the Kalman filtering system is linear and it is only suitable for reducing the interference of Gaussian noise, it has poor processing ability in complex and changeable road environments and cannot obtain good tracking results.

Table 6.2 Hungarian assignment algorithm process

Step 1	Subtract the minimum value of this row from each row of matrix $\boldsymbol{C}$, so that each row must have a zero element, and get matrix $\boldsymbol{C_1}$
Step 2	Subtract the minimum value of this column from each column of matrix $\boldsymbol{C_1}$, so that each column must have a zero element, and the column with zero element does not need to be transformed, and the matrix $\boldsymbol{C_2}$ is obtained
Step 3	Connect all zero elements in matrix $\boldsymbol{C_2}$ using the fewest row and column lines to get matrix $\boldsymbol{C_3}$. If the number of lines is equal to the matrix dimension, go directly to step 5
Step 4	Find the unconnected elements of the row and column lines in the matrix $\boldsymbol{C_3}$, subtract the minimum element from these elements, and add the minimum value at the intersection of the row and column lines to get the matrix $\boldsymbol{C_4}$, let $\boldsymbol{C_2} = \boldsymbol{C_4}$, go back to step 2
Step 5	Starting from the row or column with the fewest zero elements, select the optimal value in this row or column, and record it as the matching result until all target assignments are completed

For nonlinear and non-Gaussian problems, particle filtering can achieve better tracking results. The particle filtering uses the state of the particle set to estimate the target probability, that is, the posterior probability of the particle is used to represent its distribution, and the expectation of the particle is calculated (instead of the integral operation), and finally the minimum variance estimate of the system state is obtained. Particle filtering can adapt to nonlinear and non-Gaussian environments, but its stability is not high and its real-time performance is poor, so it cannot be directly applied to multi-target tracking.

In this section, a KF-PF joint filtering algorithm combining Kalman filtering and particle filtering is proposed to deal with the complex and changeable road environment.

1. Kalman filtering

The Kalman filtering process occupies very little system memory, and has the characteristics of fast speed and easy implementation, and is widely used in radar target tracking systems. The Kalman filtering is based on a linear system model. The input terminal receives the measurement data from the target detection, and is processed by the Kalman filter system to output the predicted data for the target state at the next moment. Kalman proposed a discrete time-domain filtering scheme under a linear state-space model, which is optimal for all linear system models. State-space models are usually described by state transition equations and observation equations:

$$\mathbf{x}_{k+1} = \boldsymbol{\Phi}_k \mathbf{x}_k + \boldsymbol{\Gamma}_k \mathbf{u}_k \tag{6.8}$$

$$\mathbf{y}_k = \mathbf{H}_k \mathbf{x}_k + \mathbf{v}_k \tag{6.9}$$

Among them, $\mathbf{x}_k$ and $\mathbf{y}_k$ are $n \times 1$ dimensional state vector and the $m \times 1$ dimensional observation vector, respectively. Both $\mathbf{u}_k$ and $\mathbf{v}_k$ are zero-mean random vectors, called process noise and observation noise, and their sizes are $s \times 1$ and $m \times 1$ respectively. $\boldsymbol{\Phi}_k$ is the state transition matrix of $n \times n$. $\boldsymbol{\Gamma}_k$ represents the process noise transition matrix of $n \times s$. $\mathbf{H}_k$ is an $m \times n$ matrix called the observation transition matrix. The statistics of the above vector are as follows:

$$\begin{aligned} &\mathbb{E}\left[\mathbf{u}_k\mathbf{u}_l^T\right] = \boldsymbol{Q}\delta_{kl}, \mathbb{E}\left[\mathbf{v}_k\mathbf{v}_l^T\right] = \boldsymbol{R}\delta_{kl} \quad \forall k, l = 0, 1, \ldots \\ &\mathbb{E}\left[\mathbf{v}_k\mathbf{x}_l^T\right] = \mathbf{0}_{m\times n}, \mathbb{E}\left[\mathbf{u}_k\mathbf{v}_l^T\right] = \mathbf{0}_{s\times m} \quad \forall k, l = 0, 1, \ldots \\ &\mathbb{E}\left[\mathbf{u}_k\mathbf{y}_l^T\right] = \mathbf{0}_{s\times m}, \quad 0 \le l \le k, \end{aligned} \tag{6.10}$$

where $\mathbf{0}_{m\times n}$ is a zero-valued matrix of size $m \times n$. The purpose of Kalman filtering is to obtain the least-square estimate $\hat{\mathbf{x}}_k$ of the state $\mathbf{x}_k$ from the observation $\mathbf{y}_l (l \le k - 1)$.

Previous researchers introduced a new random variable $\tilde{\mathbf{z}}_k$, and with this random variable derived the Kalman filter recursive equation for the state-space model described by Eqs. (6.8) and (6.9):

$$\tilde{\mathbf{z}}_k = \mathbf{y}_k - \boldsymbol{H}_k\hat{\mathbf{x}}_k \tag{6.11a}$$

$$\boldsymbol{K}_k = \boldsymbol{P}_k^{\mathbf{x}}\boldsymbol{H}_k^T\left(\boldsymbol{H}_k\boldsymbol{P}_k^{\mathbf{x}}\boldsymbol{H}_k^T + \boldsymbol{R}\right)^{-1} \tag{6.11b}$$

$$\hat{\mathbf{x}}_{k+1} = \boldsymbol{\Phi}_k\hat{\mathbf{x}}_k + \boldsymbol{\Phi}_k\mathbf{K}_k\tilde{\mathbf{z}}_k \tag{6.11c}$$

$$\boldsymbol{P}_{k+1}^{\mathbf{x}} = \boldsymbol{\Phi}_k(\mathbf{I} - \mathbf{K}_k\mathbf{H}_k)\boldsymbol{P}_k^{\mathbf{x}}\boldsymbol{\Phi}_k^T + \boldsymbol{\Gamma}_k\boldsymbol{Q}\boldsymbol{\Gamma}_k^T \tag{6.11d}$$

where the random process $\tilde{\mathbf{z}}_k$ is a discrete Gaussian random process that is used to compute the update of the least-square estimate $\hat{\mathbf{x}}_k$. $\mathbf{K}_k$ is the Kalman gain matrix and $\mathbf{P}_k^{\mathbf{x}}$ is the estimation error covariance matrix at time k. Under the Kalman filter framework, $\mathbf{u}_k$ and $\mathbf{v}_k$ are usually considered as Gaussian linear white noise, therefore, among all linear and nonlinear filters, the Kalman filter is the optimal filter in the mean square error sense.

2. Particle Filtering Based on Kalman Importance Sampling

Based on the KF-PF joint filtering idea, Kalman filtering and particle filtering can be regarded as the application of Bayesian estimation, and the unity is achieved in the sense of probability. Therefore, the particle filtering can be optimized by taking advantage of the small computational cost of Kalman filtering, and then the PF-KIS (with the particle filtering as the framework, Kalman filtering embedded in the particle filtering framework) algorithm is used for real-time multi-target tracking.

This section introduces importance sampling and uses the pre-estimation results of Kalman filtering to guide the generation of importance distributions. After the Kalman filtering pre-estimation is completed, the particle state generated by the importance sampling is closer to the real state of the target, thereby increasing the number of effective particles and improving the estimation accuracy. The estimation error can be expressed as

$$\left|I - \hat{I}\right| \leq \frac{\lambda_\alpha \sigma}{\sqrt{m}} \tag{6.12}$$

In the formula, I is the estimated value of the target, $\hat{I}$ is the estimated value obtained by particle filtering, λ_α is the coefficient, m is the number of particles, and σ is the estimated variance. The number of particles has a positive correlation with the calculation time. Reducing the number of particles under the premise of ensuring the accuracy of the algorithm can shorten the algorithm time and ensure real-time performance.

The algorithm flow is as follows.

(1) Initialize the particles to obtain the particle set $\{\boldsymbol{x}_0^i,\ i = 1,\ 2, \ldots, N\}$ and the mean $\hat{\boldsymbol{x}}_0^i$ and variance $\boldsymbol{P}_0^i$ of the particle set.

(2) Establish a state-space model.

$$\begin{cases} \boldsymbol{x}_k = f(\boldsymbol{x}_{k-1}) + \boldsymbol{u}_k \\ z_k = h(\boldsymbol{x}_k) + \boldsymbol{v}_k \end{cases} \tag{6.13}$$

Note that, unlike linear state-space models (6.8) and (6.9), Eq. (6.13) are representations of nonlinear state-space models.

(3) Use Kalman filtering to update the sampled particles, perform state prediction and update, and obtain the updated $\hat{\boldsymbol{x}}_k^i$ and $\boldsymbol{P}_k^i$.

(4) Use the Kalman filtering result to guide the importance sampling, that is, according to the sampling function F_s, the updated particle set is obtained as

$$\tilde{\boldsymbol{x}}_k^i = F_s(\hat{\boldsymbol{x}}_k^i,\ \boldsymbol{P}_k^i) \tag{6.14}$$

(5) Calculate the weight $\boldsymbol{w}_k^i$, and recalculate the target state and variance estimation

$$\boldsymbol{w}_k^i = \frac{\frac{p(z_k^i|\tilde{\boldsymbol{x}}_k^i)p(\tilde{\boldsymbol{x}}_k^i|\boldsymbol{x}_{k-1}^i)}{q(\boldsymbol{x}_k^i|\boldsymbol{x}_{k-1}^i, z_k)}}{\sum_{i=1}^{N} \frac{p(z_k^i|\tilde{\boldsymbol{x}}_k^i)p(\tilde{\boldsymbol{x}}_k^i|\boldsymbol{x}_{k-1}^i)}{q(\boldsymbol{x}_k^i|\boldsymbol{x}_{k-1}^i, z_k)}} \tag{6.15}$$

$$\boldsymbol{x}_k^i = \sum_{i=1}^{N} \boldsymbol{w}_k^i \tilde{\boldsymbol{x}}_k^i \tag{6.16}$$

$$\boldsymbol{P}_k^i = \sum_{i=1}^{N} \boldsymbol{w}_k^i (\tilde{\boldsymbol{x}}_k^i - \hat{\boldsymbol{x}}_k^i)(\tilde{\boldsymbol{x}}_k^i - \hat{\boldsymbol{x}}_k^i)^{\mathrm{T}} \tag{6.17}$$

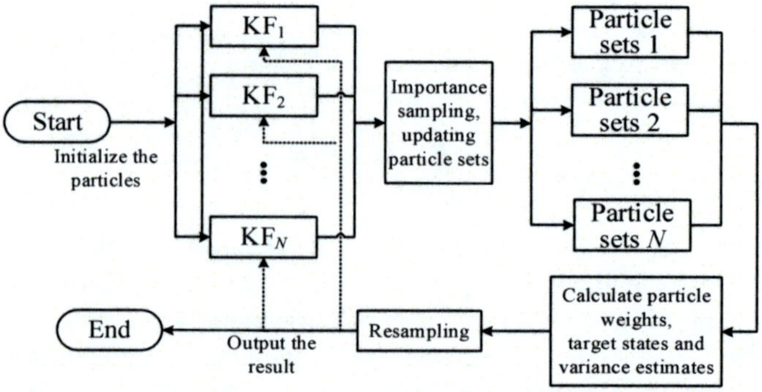

Fig. 6.5 Algorithm flow

(6) Resampling is performed using the resampling rule obtained from the vehicle motion model, the rule is

$$(R', V', N) = f(R, V, \text{SNR}) \tag{6.18}$$

In the formula, R and V represent the boundary values of the distance-velocity model, which are used to divide the associated range of the current effective target speed and distance, and resample the particles outside the range. SNR represents the signal-to-noise ratio status of a valid target. R' and $V'(R' > R,\ V' > V)$ represent the generation range of the resampled particles in the distance dimension and the velocity dimension, and the resampled particles are randomly distributed in this range. N represents the number of resampled particles. The values of R', V', and N vary with SNR.

(7) Output the target state estimate $\boldsymbol{x}_k^i$ in step (5)

The algorithm flow is shown in Fig. 6.5.

6.1.3 Experimental Comparison and Drive Test Results

The carrier of the above multi-target tracking algorithm is the millimeter-wave automobile collision avoidance radar, and the test scene is a three-lane expressway. The radar collects the data of the target vehicle ahead and evaluates the tracking effect of the proposed algorithm.

1. Track association effect

Table 6.3 compares the performance of the NN + Hungarian assignment algorithm, the JPDA algorithm, and the MHT algorithm.

As can be seen from Table 6.3, in the actual drive test scenario, the effective association rates of the three association algorithms are not significantly different,

Table 6.3 Comparison results

Algorithm	Number of targets	Processing time (ms)	Association correct rate (%)
JPDA	8	9.4	96.7
MHT	8	10.2	96.9
NN + Hungarian assignment	8	4.8	95.6
JPDA	10	15.3	96.2
MHT	10	16.2	96.6
NN + Hungarian assignment	10	6.1	94.7

but the processing time of the NN + Hungarian assignment algorithm is significantly shorter than that of the JPDA algorithm and the MHT algorithm. In addition, when the number of targets increases, the algorithm proposed in this paper can still maintain good real-time performance.

Next, perform multi-frame statistics on the 4 scenarios, and evaluate the track association algorithm through indicators such as the number of correct associations, the number of incorrect associations, the number of unassociated associations, and the effective association rate. The results are shown in Table 6.4.

Unassociated indicates a situation where the assigned target is too different from the track to be associated. Effective association rate = Number of correct associations/(Number of correct associations + Number of false associations).

The average effective association rates of the four scenarios in Table 6.4 are all around 95%, indicating that the track association algorithm can effectively match targets and tracks in multi-target scenarios, and has good practical value.

2. Moving target tracking algorithm effect

Taking the driving scene of changing speed and lane as the test scene of the filtering algorithm, the PF-KIS algorithm is compared with the commonly used KF algorithm and PF algorithm.

Table 6.4 Performance comparison of track association algorithms

	Scene 1	Scene 2	Scene 3	Scene 4
Number of periods	50	100	100	100
Number of targets	10	8	9	9
Number of correct associations	394	687	678	764
Number of false associations	22	31	35	41
Unassociated numbers	64	82	87	95
Effective association rate	94.7	95.6	95.1	94.7

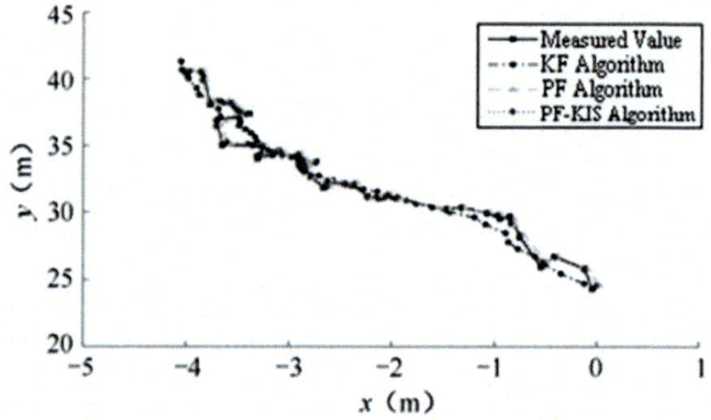

Fig. 6.6 Comparison of tracking effects of KF algorithm, PF algorithm, and PF-KIS algorithm

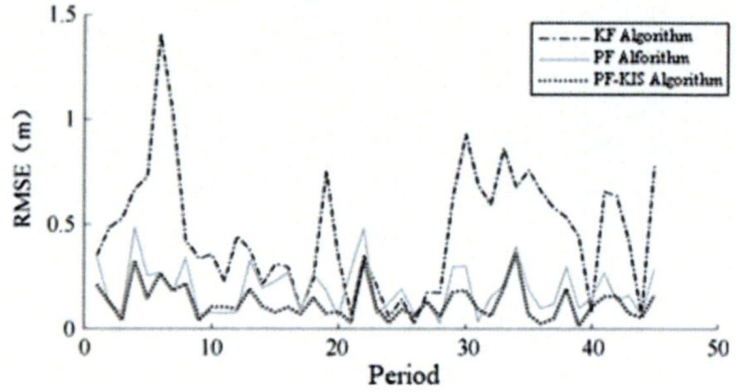

Fig. 6.7 Comparison of tracking errors of KF algorithm, PF algorithm, and PF-KIS algorithm

First, simulate the irregular motion trajectory of the vehicle accelerating to the left front. Next, the vehicle is tracked using the algorithm described above based on the simulated data. Finally, the performance of each algorithm is evaluated according to the tracking results. Figure 6.6 shows the difference in tracking effect between the KF algorithm, the PF algorithm, and the PF-KIS algorithm. The tracking errors of the KF algorithm, PF algorithm, and PF-KIS algorithm are shown in Fig. 6.7.

As can be seen from Fig. 6.7, in many cases, the tracking error of the KF algorithm is greater than 0.5 m, and the accuracy is low, which cannot meet the actual needs. In contrast, the average error of the PF algorithm and the PF-KIS algorithm is much smaller than that of the KF algorithm, and the average error of the PF-KIS algorithm is slightly smaller than that of the PF algorithm. The results show that the effect of PF-KIS algorithm is better than that of KF algorithm and PF algorithm.

Next, the PF algorithm and the PF-KIS algorithm are further compared. The processing time and tracking error of the PF algorithm and the PF-KIS algorithm under different target numbers are shown in Table 6.5.

As can be seen from Table 6.5, when the number of targets is the same, the processing time of the PF-KIS algorithm is significantly shorter than that of the PF algorithm. The tracking error of the PF-KIS algorithm when there are 50 targets is not much different from that of the PF algorithm when there are 150 targets, and the processing time of the former is about 1/5 of the latter. Therefore, the PF-KIS algorithm has higher tracking accuracy and stronger real-time performance.

Table 6.5 Comparison results of processing time and tracking error between PF algorithm and PF-KIS algorithm under different target numbers

Algorithm	Number of targets	Processing time (ms)	Tracking error (m)
PF	50	0.97	0.4957
PF−KIS	50	0.66	0.3417
PF	100	1.93	0.4141
PF−KIS	100	1.08	0.2832
PF	150	3.19	0.3225
PF−KIS	150	1.94	0.2245

3. Actual drive test results

Test equipment: PC, radar, camera, etc.

Test location: a three-lane highway.

During the actual test, the host computer simultaneously displays the screen, the speed-distance two-dimensional graph, the target detected in the current period, and the track with high confidence. The test interface is shown in Fig. 6.8.

The radar can distinguish the target well and obtain the state information of each target accurately. At the same time, the time cost of realizing the whole process

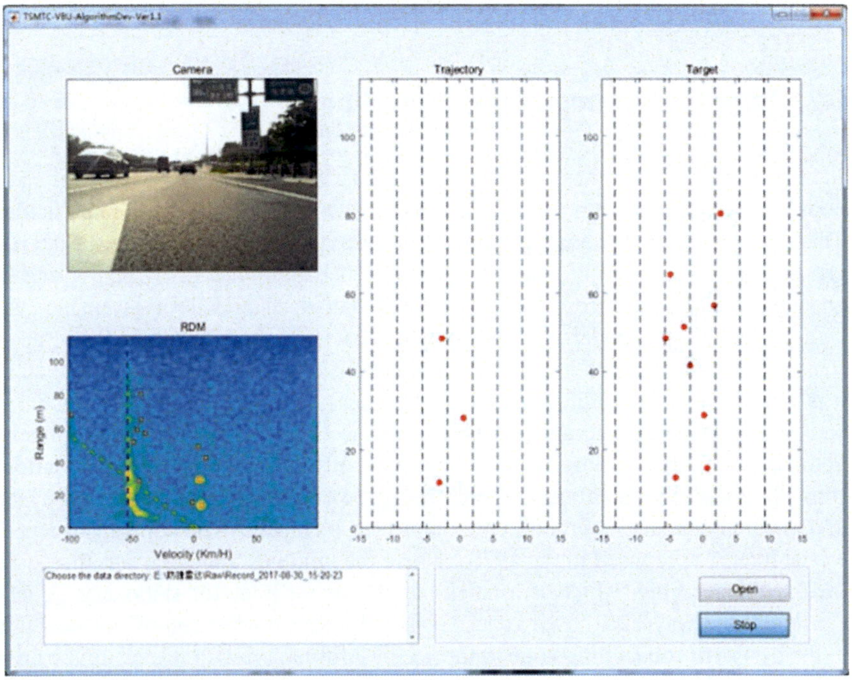

Fig. 6.8 Host computer interface

meets the real-time requirements of forward collision avoidance radar, and achieves the expected effect.

In the experiment, the road test results of multiple scenarios are analyzed, mainly including straight-line driving scenarios, lane-changing driving scenarios, curved driving scenarios and multi-target scenarios. Set the number of targets to 50, and record the track of the target tracked by the PF-KIS algorithm within a certain period of time. Figure 6.9 shows the single target drive test results in different scenarios.

Figure 6.9 shows the tracking effect of a single target in a straight-line driving scenario, a lane-changing driving scenario, and a curved driving scenario. Since the target reflection surface is mainly lateral, there is a certain lateral fluctuation during tracking, which in turn causes the tracking result to drift in the straight-line driving scene. In contrast, in the lane-change driving scene and the curve driving scene, the lateral drift of the track is not obvious due to the lateral displacement of the vehicle itself. In the above three scenarios, the target track is continuous and stable, and has good scene adaptability.

Figure 6.10 shows the drive test results in a multi-target scenario.

As can be seen from Fig. 6.10, in the multi-target scene, the proposed algorithm can distinguish targets well, track the state of each moving target accurately and stably, and has good multi-target tracking performance. In conclusion, the PF-KIS algorithm has become a necessary condition to achieve the goal of security early warning.

6.2 Improved Kalman Filter Algorithm

In engineering applications, the design of robust filters has been one of the most popular topics in modern radar systems and MIMO-based radar system in particular. There are two main reasons why this has occurred. First, a radar signal of interest can be corrupted naturally from other interference and noise or by clutter and so on, which results in uncertainty in assumed statistical models of given received signals. Second, the models of measurement system can be affected by the setup of transmitters and receivers in practical applications of radar systems. Unfortunately, it is impossible or prohibitively expensive to design an optimal filter by obtaining and/or understanding accurate models in the real world, which makes some nominally optimal filters suffer significant degradations in performance even if small deviations from the assumed models occur. Therefore, a robust filter can be considered as a kind of optimal filter under some conditions of uncertainty models. The seminal concept of robust filters is proposed in the 1970s, where the minimax approach was discussed for the Wiener filter, which is a batch-based optimal filter for stationary signals. Currently, the Bayesian-based robust approach has attracted significant attention in the design of robust filters due to the use of information with respect to the prior distribution of models in providing more accurate knowledge on the statistical model.

It is known to all that classical Kalman filtering involves an optimal filter in terms of the Gaussian and linear assumptions of dynamic systems. It has been well

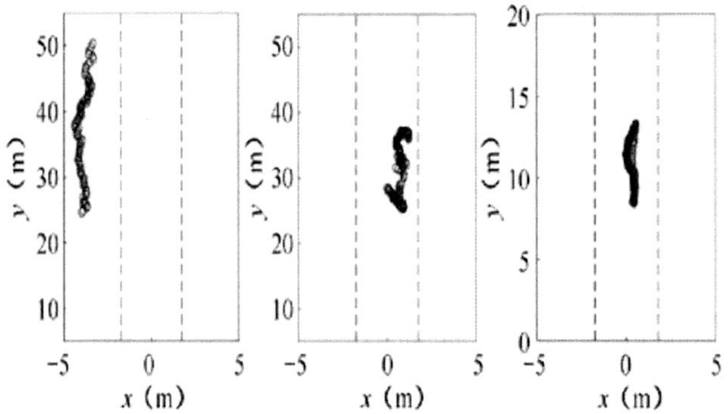

(a) Straight-line driving scenario

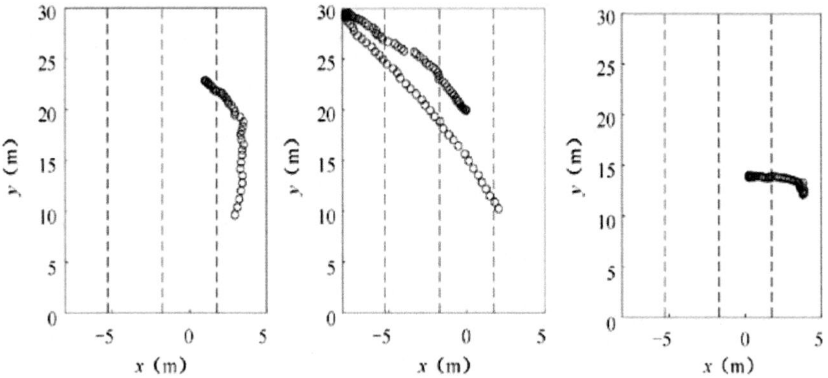

(b) Lane-changing driving scenario

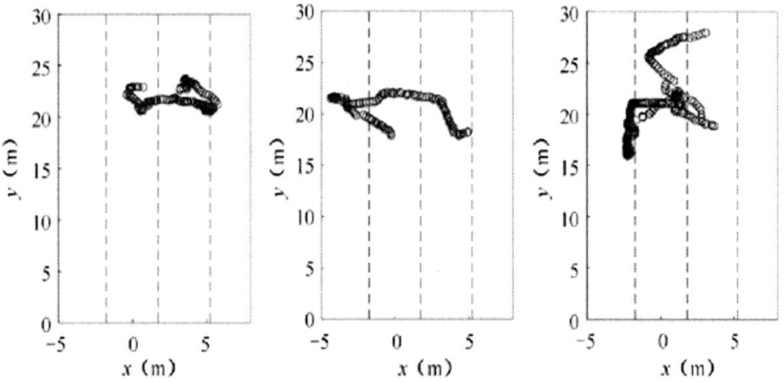

(c) Curved driving scenario

Fig. 6.9 Single-target drive test results in different scenarios

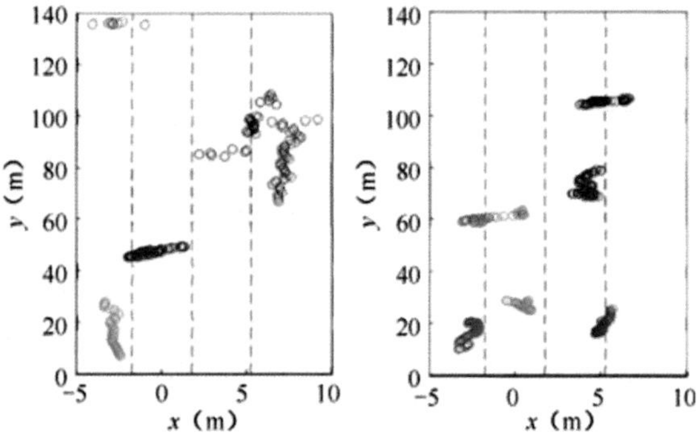

Fig. 6.10 Drive test results in a multi-target scenario

established in many fields such as communication, navigation, radar and control. In addition, more and more Kalman filtering-related technologies are applied to the localization and tracking of moving agents. However, it also suffers from prominent shortcomings, for example, the good performance lies in completely grasping the statistical models of noise and the state of signal. As a result, the problem of designing a robust Kalman filter in practical applications where the knowledge of noise distribution is missing or inaccurate is a big challenge for both researchers and developers. Several solutions have been proposed by previous researchers, including so-called adaptive Kalman filtering. These filters can estimate both signal state and noise simultaneously, and this approach works well when there is a large amount of data used to obtain some accurate performance over the entire estimation period. Later, minimax-based Kalman filters and finite impulse response Kalman filters were proposed.

To address the tracking problem when the noise model is inaccurate, this section introduces a Bayesian robust Kalman filter with inaccurate noise second-order statistics.

6.2.1 Bayesian Robust Kalman Filter Based on Posterior Noise Statistics (KFPNS)

Thanks to the pioneering work of Bode, Shannon, etc., the concept of innovation process is widely used in the derivation of Kalman recursive equations. In this subsection, the recursive equations of KFPNS are derived based on Bayesian theory. Compared with the recursive equations of a classical Kalman filter, it only needs to calculate the posterior noise statistics, which are obtained by variational distribution. The variational distribution is an approximate distribution of the real posterior noise distribution.

1. Introduction to Bayesian Robust Filters

In engineering, the Bayesian criterion and minimax criterion (instead of the Bayesian method when there is no prior knowledge) are usually used to design robust filters. The meaning of a Bayesian criterion is as follows: The average cost is the least. In the context of state estimation, the average cost is generally reflected by the mean-square error (MSE); that is

$$C(\mathbf{x}_k, \psi(\mathbf{y}_k)) = \mathbb{E}\left[(\mathbf{x}_k - \psi(\mathbf{y}_k))^T(\mathbf{x}_k - \psi(\mathbf{y}_k))\right] \tag{6.19}$$

According to the Bayesian criterion, the optimal filter can be obtained by minimizing the cost function:

$$\hat{\psi}(\mathbf{y}_k) = \arg\min_{\psi\in\Psi} C(\mathbf{x}_k, \psi(\mathbf{y}_k)) \tag{6.20}$$

where Ψ is the possible class of filters. The filter that satisfies the above formula is the minimum mean square error filter.

When all knowledge of the model cannot be obtained, it can be assumed that the model is controlled by an unknown parameter $\boldsymbol{\theta} = [\theta_1, \ldots, \theta_l]$, $\boldsymbol{\theta} \in \Theta$, where Θ is the set of all possible parameters, called the uncertain class. Fortunately, these parameters can be obtained by solving the corresponding posterior distribution, thus, the Bayesian robust filter can be expressed as follows:

$$\psi^{\Theta}(\mathbf{y}_k) = \arg\min_{\psi\in\Psi} \mathbb{E}_{\boldsymbol{\theta}}\left[C_{\boldsymbol{\theta}}(\mathbf{x}_k, \psi(\mathbf{y}_k))\,|\mathcal{Y}_k\right] \tag{6.21}$$

where the expectation is taken relative to the posterior distribution $p(\boldsymbol{\theta}|\mathcal{Y}_k)$, and $C_{\boldsymbol{\theta}}(\cdot)$ is the cost function relative to the unknown parameter $\boldsymbol{\theta}$. Some researchers have proposed that the average cost of the filter relative to the uncertain class is equal to the error of applying the filter to effective information. Therefore, designing a robust filter for uncertain models can be transformed into designing a Bayesian robust filter using posterior effective information.

2. Derivation of KFPNS Recursive Equation

Now assume that the second-order statistical knowledge of process noise and observation noise is unknown and that it is determined by the unknown parameters α_1 and α_2. Here, the covariance matrices of these two noises are as follows.

$$\mathbb{E}\left[\mathbf{u}_k^{\alpha_1}\left(\mathbf{u}_l^{\alpha_1}\right)^T\right] = \boldsymbol{Q}^{\alpha_1}\delta_{kl} \tag{6.22}$$

$$\mathbb{E}\left[\mathbf{v}_k^{\alpha_2}\left(\mathbf{v}_l^{\alpha_2}\right)^T\right] = \boldsymbol{R}^{\alpha_2}\delta_{kl} \tag{6.23}$$

If α_1 and α_2 are statistically independent, then $p(\boldsymbol{\alpha}) = p(\alpha_1)p(\alpha_2)$ stands for the prior distribution. The state-space model can be parameterized as follows.

$$\mathbf{x}_{k+1}^{\alpha_1} = \boldsymbol{\Phi}_k \mathbf{x}_k^{\alpha_1} + \boldsymbol{\Gamma}_k \mathbf{u}_k^{\alpha_1} \tag{6.24}$$

$$\mathbf{y}_k^{\boldsymbol{\alpha}} = \boldsymbol{H}_k \mathbf{x}_k^{\alpha_1} + \mathbf{v}_k^{\alpha_2} \tag{6.25}$$

With $\mathcal{Y}_k = \{\mathbf{y}_0, \ldots, \mathbf{y}_k\}$ as the observation sequence and analysis based on (6.21), the linear filtering function that satisfies the definition of a Bayesian robust filter is as follows:

$$\hat{\mathbf{x}}_k^{\boldsymbol{\alpha}} = \sum_{l \le k-1} \boldsymbol{\Omega}_{k,l}^{\Theta} \mathbf{y}_l^{\boldsymbol{\alpha}} \tag{6.26}$$

such that the following is the case:

$$\boldsymbol{\Omega}_{k,l}^{\Theta} = \arg \min_{\boldsymbol{\Omega}_{k,l} \in \mathfrak{J}} \mathbb{E}_{\boldsymbol{\alpha}} \left[\mathbb{E}\left[\left(\mathbf{x}_k^{\alpha_1} - \sum_{l \le k-1} \boldsymbol{\Omega}_{k,l} \mathbf{y}_l^{\boldsymbol{\alpha}} \right)^T \left(\mathbf{x}_k^{\alpha_1} - \sum_{l \le k-1} \boldsymbol{\Omega}_{k,l} \mathbf{y}_l^{\boldsymbol{\alpha}} \right) \right] \middle| \mathcal{Y}_{k-1} \right] \tag{6.27}$$

where $\mathfrak{J}$ is the vector space of all $n \times m$ matrix-valued functions, and $\boldsymbol{\Omega}_{k,l} \in \mathfrak{J}$ is a mapping $\boldsymbol{\Omega}_{k,l} : \mathrm{N} \times \mathrm{N} \to \mathbb{R}^{n \times m}$ such that $\sum_{k=1}^{\infty} \sum_{l=1}^{\infty} \left\| \boldsymbol{\Omega}_{k,l} \right\|_2 < \infty$, $\|\cdot\|_2$ being the L_2 norm.

The derivation of KFPNS recursive equations depends on the following theorem, definition and lemma.

Theorem 1. (Bayesian Orthogonality Principle) *If the weight function* $\boldsymbol{\Omega}_{k,l}^{\Theta}$ *of a linear filter satisfies* (6.27), *then the* $\hat{\mathbf{x}}_k^{\boldsymbol{\alpha}}$ *obtained in* (6.26) *is called the optimal Bayesian least-squares estimation if and only if the following is the case.*

$$\mathbb{E}_{\boldsymbol{\alpha}}\left[\mathbb{E}\left[\left(\mathbf{x}_k^{\alpha_1} - \hat{\mathbf{x}}_k^{\boldsymbol{\alpha}} \right) \left(\mathbf{y}_l^{\boldsymbol{\alpha}} \right)^T \right] \middle| \mathcal{Y}_{k-1} \right] = \mathbf{0}_{n \times m} \; \forall l \le k-1 \tag{6.28}$$

Definition 1. (*Bayesian innovation process*) According to the state-space model described in (6.24) and (6.25), if $\hat{\mathbf{x}}_k^{\boldsymbol{\alpha}}$ is the least-square estimate of $\mathbf{x}_k^{\alpha_1}$, then the stochastic process of the following:

$$\tilde{\mathbf{z}}_k^{\boldsymbol{\alpha}} = \mathbf{y}_k^{\boldsymbol{\alpha}} - \boldsymbol{H}_k \hat{\mathbf{x}}_k^{\boldsymbol{\alpha}} \tag{6.29}$$

is a zero-mean process and $\forall l, l' \le k-1$; the following is obtained.

$$\mathbb{E}_{\boldsymbol{\alpha}}\left[\mathbb{E}\left[\tilde{\mathbf{z}}_l^{\boldsymbol{\alpha}} \left(\tilde{\mathbf{z}}_{l'}^{\boldsymbol{\alpha}} \right)^T \right] \middle| \mathcal{Y}_{k-1} \right] = \mathbb{E}_{\boldsymbol{\alpha}}\left[\boldsymbol{H}_l \boldsymbol{P}_k^{\mathbf{x},\boldsymbol{\alpha}} \boldsymbol{H}_l^T + \boldsymbol{R}^{\alpha_2} \middle| \mathcal{Y}_{k-1} \right] \delta_{ll'} \tag{6.30}$$

Lemma 1. (Bayesian information equivalence) *The Bayesian least-squares estimation obtained based on* $\tilde{\mathbf{z}}_k^{\alpha}$ *is equivalent to the Bayesian least-squares estimation calculated based on* $\mathbf{y}_k^{\alpha}$*; in other words, we have the following.*

$$\mathbb{E}_{\alpha}\Big[\mathbb{E}\Big[\big(\mathbf{x}_k^{\alpha_1} - \hat{\mathbf{x}}_k^{\alpha}\big)\big(\tilde{\mathbf{z}}_k^{\alpha}\big)^T\Big]\Big|\mathcal{Y}_{k-1}\Big] = \mathbf{0}_{n\times m} \tag{6.31}$$

Then, (6.26) can be rewritten as follows.

$$\hat{\mathbf{x}}_k^{\alpha} = \sum_{l\le k-1} \boldsymbol{\Omega}_{k,l}^{\Theta}\tilde{\mathbf{z}}_l^{\alpha} \tag{6.32}$$

By substituting it into (6.31) and listing the weight function alone on the left side of the equation, the following is obtained.

$$\boldsymbol{\Omega}_{k,l}^{\Theta} = \mathbb{E}_{\alpha}\Big[\mathbb{E}\Big[\mathbf{x}_k^{\alpha_1}\big(\tilde{\mathbf{z}}_l^{\alpha}\big)^T\Big]\Big|\mathcal{Y}_{k-1}\Big]\mathbb{E}_{\alpha}^{-1}\big[\boldsymbol{H}_l\boldsymbol{P}_l^{\mathbf{x},\alpha}\boldsymbol{H}_l^T + \boldsymbol{R}^{\alpha_2}\big|\mathcal{Y}_{k-1}\big] \tag{6.33}$$

In this case, the linear function of Bayesian robust filter based on posterior noise statistics is as follows.

$$\hat{\mathbf{x}}_k^{\alpha} = \sum_{l\le k-1}\mathbb{E}_{\alpha}\Big[\mathbb{E}\Big[\mathbf{x}_k^{\alpha_1}\big(\tilde{\mathbf{z}}_l^{\alpha}\big)^T\Big]\Big|\mathcal{Y}_{k-1}\Big]\mathbb{E}_{\alpha}^{-1}\big[\boldsymbol{H}_l\boldsymbol{P}_l^{\mathbf{x},\alpha}\boldsymbol{H}_l^T + \boldsymbol{R}^{\alpha_2}\big|\mathcal{Y}_{k-1}\big]\tilde{\mathbf{z}}_l^{\alpha} \tag{6.34}$$

Thus, the state update equation is as follows:

$$\hat{\mathbf{x}}_{k+1}^{\alpha} = \boldsymbol{\Phi}_k\hat{\mathbf{x}}_k^{\alpha} + \boldsymbol{\Phi}_k\mathbf{K}_k^{\Theta}\tilde{\mathbf{z}}_k^{\alpha} \tag{6.35}$$

where the following is the case.

$$\boldsymbol{K}_k^{\Theta'} = \mathbb{E}_{\alpha}\big[\boldsymbol{P}_k^{\mathbf{x},\alpha}\big|\mathcal{Y}_{k-1}\big]\boldsymbol{H}_k^T\mathbb{E}_{\alpha}^{-1}\big[\boldsymbol{H}_k\boldsymbol{P}_k^{\mathbf{x},\alpha}\boldsymbol{H}_k^T + \boldsymbol{R}^{\alpha_2}\big|\mathcal{Y}_{k-1}\big] \tag{6.36}$$

Next, let $\tilde{\mathbf{x}}_k^{\alpha} = \mathbf{x}_k^{\alpha_1} - \hat{\mathbf{x}}_k^{\alpha}$ denote the estimation error at time k. After some mathematical operations, $\tilde{\mathbf{x}}_{k+1}^{\alpha}$ can be expanded into the following.

$$\tilde{\mathbf{x}}_{k+1}^{\alpha} = \boldsymbol{\Phi}_k\Big(\boldsymbol{I} - \boldsymbol{K}_k^{\Theta'}\boldsymbol{H}_k\Big)\tilde{\mathbf{x}}_k^{\alpha} + \boldsymbol{\Gamma}_k\mathbf{u}_k^{\alpha_1} - \boldsymbol{\Phi}_k\boldsymbol{K}_k^{\Theta'}\mathbf{v}_k^{\alpha_2} \tag{6.37}$$

An update equation to estimation error covariance matrix can be found based on the above equation, which is described below.

$$\mathbb{E}_{\alpha}\big[\boldsymbol{P}_{k+1}^{\mathbf{x},\alpha}|\mathcal{Y}_k\big] = \boldsymbol{\Phi}_k\Big(\mathbf{I} - \mathbf{K}_k^{\Theta'}\mathbf{H}_k\Big)\mathbb{E}_{\alpha}\big[\boldsymbol{P}_k^{\mathbf{x},\alpha}\big|\mathcal{Y}_k\big]\boldsymbol{\Phi}_k^T + \boldsymbol{\Gamma}_k\mathbb{E}_{\alpha}\big[\boldsymbol{Q}^{\alpha_1}|\mathcal{Y}_k\big]\boldsymbol{\Gamma}_k^T \tag{6.38}$$

This completes the construction of KFPNS. Note that in (6.36) and (6.38), the observation sequences corresponding to the estimation error covariance matrix are different, which are $\mathbb{E}_{\alpha}\big[\mathbf{P}_k^{\mathbf{x},\alpha}\big|\mathcal{Y}_{k-1}\big]$ and $\mathbb{E}_{\alpha}\big[\mathbf{P}_k^{\mathbf{x},\alpha}\big|\mathcal{Y}_k\big]$, respectively, while the one generated by the last iteration of the recursive equation should be $\mathbb{E}_{\boldsymbol{\alpha}}\left[\mathbf{P}_k^{\mathbf{x},\boldsymbol{\alpha}}|\mathcal{Y}_{k-1}\right]$. For this problem, the first solution is to iterate from the beginning; that is, first calculate $\boldsymbol{K}_0^{\Theta'}$ with $\mathbb{E}_{\boldsymbol{\alpha}}\left[\mathbf{P}_0^{\mathbf{x},\boldsymbol{\alpha}}|\mathcal{Y}_{-1}\right]$ and $\mathbb{E}_{\boldsymbol{\alpha}}\left[\boldsymbol{R}^{\alpha_2}|\mathcal{Y}_k\right]$, then use $\boldsymbol{K}_0^{\Theta'}$ and $\mathbb{E}_{\boldsymbol{\alpha}}[\mathbf{Q}^{\alpha_1}|\mathcal{Y}_k]$ to compute $\mathbb{E}_{\boldsymbol{\alpha}}\left[\mathbf{P}_1^{\mathbf{x},\boldsymbol{\alpha}}|\mathcal{Y}_k\right]$ and repeat the process until obtain $\mathbb{E}_{\boldsymbol{\alpha}}\left[\mathbf{P}_k^{\mathbf{x},\boldsymbol{\alpha}}|\mathcal{Y}_k\right]$. However, this strategy requires reiteration from the origin after updating posterior noise statistics, which makes the algorithm heavily loaded. In order to improve the efficiency of the algorithm, it is also feasible to assume $\mathbb{E}_{\boldsymbol{\alpha}}\left[\mathbf{P}_k^{\mathbf{x},\boldsymbol{\alpha}}|\mathcal{Y}_{k-1}\right] \approx \mathbb{E}_{\boldsymbol{\alpha}}\left[\mathbf{P}_k^{\mathbf{x},\boldsymbol{\alpha}}|\mathcal{Y}_k\right]$.

Table 6.6 reports the difference between these two recursive equations. KFPNS replaces $\mathbf{K}_k$, $\mathbf{P}_k$, $\mathbf{Q}$ and $\mathbf{R}$ in the classical Kalman filter with $\mathbf{K}_k^{\Theta'}$, $\mathbb{E}_{\boldsymbol{\alpha}}\left[\mathbf{P}_k^{\mathbf{x},\boldsymbol{\alpha}}|\mathcal{Y}_{k-1}\right]$, $\mathbb{E}_{\boldsymbol{\alpha}}[\mathbf{Q}^{\alpha_1}|\mathcal{Y}_k]$ and $\mathbb{E}_{\boldsymbol{\alpha}}[\mathbf{R}^{\alpha_2}|\mathcal{Y}_k]$, but their recursive structures are the same. Figure 6.11 is the overall framework of KFPNS. The key is that KFPNS updates the state based on posterior noise statistics. Therefore, KFPNS is an improved version of the classical Kalman filter concerning posterior noise distribution.

3. Calculation of posterior effective noise statistics

The execution of KFPNS requires additional computation of $\mathbb{E}_{\boldsymbol{\alpha}}[\mathbf{Q}^{\alpha_1}|\mathcal{Y}_k]$ and $\mathbb{E}_{\boldsymbol{\alpha}}[\mathbf{R}^{\alpha_2}|\mathcal{Y}_k]$, and the core of obtaining these two conditional expectations lies in the posterior distribution $\mathbb{E}_{\boldsymbol{\alpha}}[\mathbf{R}^{\alpha_2}|\mathcal{Y}_k]$, which is unknown. Therefore, the goal is to solve the probability density function of the unknown parameter $\boldsymbol{\alpha}$ under the condition of given observation sequence $\mathcal{Y}_k$. Although $p(\boldsymbol{\alpha}|\mathcal{Y}_k) \propto f(\mathcal{Y}_k|\boldsymbol{\alpha})p(\boldsymbol{\alpha})$, it is difficult to directly acquire a closed-form solution by relying on the prior distribution. However, a known simple distribution can be used to approximate a complex distribution, and by limiting the type of approximate distribution, a locally optimal approximate posterior distribution can be achieved. Unless otherwise specified, the distributions presented in this paper are Gaussian.

Consider a simple variational distribution, which belongs to the following exponential family:

$$q(\alpha; \mathbf{e}) = g(\boldsymbol{\alpha}) \exp\left\{\mathbf{e}^T t(\boldsymbol{\alpha}) - A(\mathbf{e})\right\} \tag{6.39}$$

where $g(\boldsymbol{\alpha})$ is the base measure, $\mathbf{e}$ is the natural parameter, $t(\boldsymbol{\alpha})$ is the sufficient statistics and $A(\mathbf{e})$ is the log-normalizer. Here, e can be considered to represent the

Table 6.6 Comparison of the recursive equations for the classical algorithm and KFPNS

Classical Kalman Filter	KFPNS
$\tilde{\mathbf{z}}_k = \mathbf{y}_k - \mathbf{H}_k\hat{\mathbf{x}}_k$	$\tilde{\mathbf{z}}_k^{\boldsymbol{\alpha}} = \mathbf{y}_k^{\boldsymbol{\alpha}} - \mathbf{H}_k\hat{\mathbf{x}}_k^{\boldsymbol{\alpha}}$
$\mathbf{K}_k = \mathbf{P}_k^{\mathbf{x}}\mathbf{H}_k^T\left(\mathbf{H}_k\mathbf{P}_k^{\mathbf{x}}\mathbf{H}_k^T + \mathbf{R}\right)^{-1}$	$\mathbf{K}_k^{\Theta'} =$ $\mathbb{E}_{\boldsymbol{\alpha}}\left[\mathbf{P}_k^{\mathbf{x},\boldsymbol{\alpha}}|\mathcal{Y}_{k-1}\right]\mathbf{H}_k^T\mathbb{E}_{\boldsymbol{\alpha}}^{-1}\left[\mathbf{H}_k\mathbf{P}_k^{\mathbf{x},\boldsymbol{\alpha}}\mathbf{H}_k^T + \mathbf{R}^{\alpha_2}|\mathcal{Y}_{k-1}\right]$
$\hat{\mathbf{x}}_{k+1} = \mathbf{\Phi}_k\hat{\mathbf{x}}_k + \mathbf{\Phi}_k\mathrm{K}_k\tilde{\mathbf{z}}_k$	$\hat{\mathbf{x}}_{k+1}^{\boldsymbol{\alpha}} = \mathbf{\Phi}_k\hat{\mathbf{x}}_k^{\boldsymbol{\alpha}} + \mathbf{\Phi}_k\mathrm{K}_k^{\Theta}\tilde{\mathbf{z}}_k^{\boldsymbol{\alpha}}$
$\mathbf{P}_{k+1}^{\mathbf{x}} = \mathbf{\Phi}_k(\mathbf{I} - \mathrm{K}_k\mathrm{H}_k)\mathbf{P}_k^{\mathbf{x}}\boldsymbol{\Phi}_k^T + \boldsymbol{\Gamma}_k\mathbf{Q}\boldsymbol{\Gamma}_k^T$	$\mathbb{E}_{\alpha}[\mathbf{P}_{k+1}^{\mathbf{x},\boldsymbol{\alpha}}|\mathcal{Y}_k] = \mathbf{\Phi}_k\left(\mathrm{I} - \mathrm{K}_k^{\Theta'}\mathrm{H}_k\right)\mathbb{E}_{\boldsymbol{\alpha}}\left[\mathbf{P}_k^{\mathbf{x},\boldsymbol{\alpha}}|\mathcal{Y}_k\right]\mathbf{\Phi}_k^T$ $+\boldsymbol{\Gamma}_k\mathbb{E}_{\boldsymbol{\alpha}}\left[\mathbf{Q}^{\alpha_1}|\mathcal{Y}_k\right]\boldsymbol{\Gamma}_k^T$

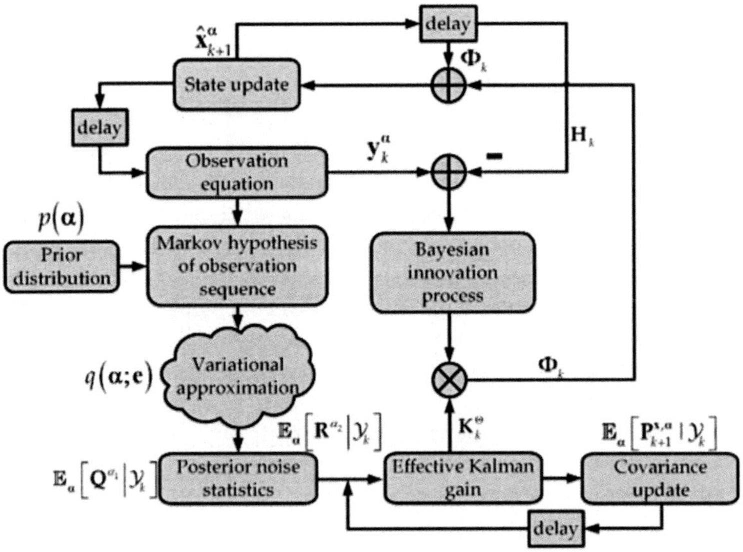

Fig. 6.11 KFPNS framework

expectations, $t(\boldsymbol{\alpha})$ can completely characterize all quantities of a probability distribution and $t(\boldsymbol{\alpha}) = \boldsymbol{\alpha}$. Take a simple multivariate Gaussian distribution $\mathcal{N}(\boldsymbol{\alpha}; \mathbf{e}, \boldsymbol{\Lambda})$ as an example ($\boldsymbol{\Lambda}$ represents the identity matrix) and transform it into an exponential family form of the following:

$$\begin{aligned}
p(\boldsymbol{\alpha};\mathbf{e},\boldsymbol{\Lambda}) &= \frac{1}{(2\pi)^{d/2}|\boldsymbol{\Lambda}|^{1/2}}\exp\left\{-\frac{1}{2}(\boldsymbol{\alpha}-\mathbf{e})^T\boldsymbol{\Lambda}^{-1}(\boldsymbol{\alpha}-\mathbf{e})\right\} \\
&= (2\pi)^{-d/2}|\boldsymbol{\Lambda}|^{1/2}\exp\left\{\mathbf{e}^T\boldsymbol{\alpha}-\frac{1}{2}\boldsymbol{\alpha}^T\boldsymbol{\Lambda}\boldsymbol{\alpha}-\frac{1}{2}\mathbf{e}^T\boldsymbol{\Lambda}\mathbf{e}\right\} \\
&= (2\pi)^{-d/2}|\boldsymbol{\Lambda}|^{1/2}\exp\left\{-\frac{1}{2}\boldsymbol{\Lambda}\boldsymbol{\alpha}^T\boldsymbol{\alpha}\right\}\exp\left\{\mathbf{e}^T\boldsymbol{\alpha}-\frac{1}{2}\boldsymbol{\Lambda}\mathbf{e}^T\mathbf{e}\right\},
\end{aligned} \tag{6.40}$$

thus, the following is obtained.

$$\begin{aligned}
g(\boldsymbol{\alpha}) &= (2\pi)^{-d/2}|\boldsymbol{\Lambda}|^{1/2}\exp\left\{-\frac{1}{2}\boldsymbol{\Lambda}\boldsymbol{\alpha}^T\boldsymbol{\alpha}\right\}, \\
t(\boldsymbol{\alpha}) &= \boldsymbol{\alpha}, \\
A(\mathbf{e}) &= \frac{1}{2}\boldsymbol{\Lambda}\mathbf{e}^T\mathbf{e}.
\end{aligned} \tag{6.41}$$

In addition, the exponential family distribution also has the following property.

$$\nabla_{\mathbf{e}}A(\mathbf{e}) = \mathbb{E}_{q(\boldsymbol{\alpha};\mathbf{e})}[t(\boldsymbol{\alpha})] \tag{6.42}$$

By using this property, the mutual transformation between expectation calculation and derivation calculation can be completed in variational inference. Therefore, most applications of variational inference use exponential family functions. More importantly, based on this simple property, a strategy of control variates can be used to reduce the variance of the Monte Carlo gradient. At the same time, the posterior distribution $p(\boldsymbol{\alpha}|\mathbf{y}_k)$ can be approximated by the variational distribution $q(\boldsymbol{\alpha}; \mathbf{e})$, and the special form of the exponential family distribution renders this approximation process traceable. More precisely, $\boldsymbol{\alpha}$ is considered to be a Gaussian distribution, which belongs to the exponential family, and its natural parameter $\mathbf{e}$ represents expectations. Therefore, the process of finding the optimal natural parameters that makes $q(\boldsymbol{\alpha}; \mathbf{e})$ the closest to $p(\boldsymbol{\alpha}|\mathbf{y}_k)$ involves calculating posterior noise statistics.

Now, suppose $q(\boldsymbol{\alpha}; \mathbf{e})$ can be decomposed into the following.

$$q(\boldsymbol{\alpha}; \mathbf{e})=q(\alpha_1; e_1)q(\alpha_2; e_2) \tag{6.43}$$

Based on the idea of variational inference, the Kullback–Leibler divergence $D_{KL}[q(\boldsymbol{\alpha}; \mathbf{e})\|p(\boldsymbol{\alpha}|\mathcal{Y}_k)]$ can be minimized to achieve the purpose of using $q(\boldsymbol{\alpha}; \mathbf{e})$ to approximate $p(\boldsymbol{\alpha}|\mathcal{Y}_k)$. This idea still cannot be realized directly, because D_{KL} contains an unknown distribution $p(\boldsymbol{\alpha}|\mathcal{Y}_k)$. Fortunately, after some logarithmic operations, minimizing D_{KL} is equivalent to maximizing the evidence lower bound objective (ELBO).

$$\mathcal{L} = \mathbb{E}_{q(\boldsymbol{\alpha};\mathbf{e})}\left[\log \frac{p(\alpha, \mathcal{Y}_k)}{q(\boldsymbol{\alpha}; \mathbf{e})}\right] \tag{6.44}$$

The principle is shown in Fig. 6.12.

Then, ELBO takes the derivative with respect to parameter $\mathbf{e}$:

$$\nabla_{\mathbf{e}}\mathcal{L} = \mathbb{E}_{q(\boldsymbol{\alpha};\mathbf{e})}[f(\boldsymbol{\alpha})] \tag{6.45}$$

where the following is the case.

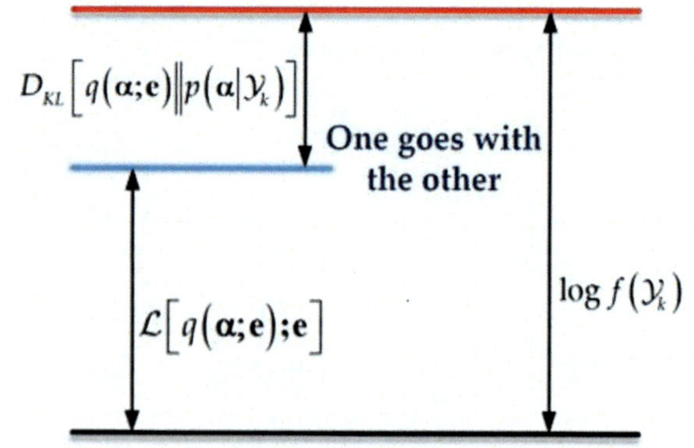

Fig. 6.12 The principle of conversion

$$f(\alpha) = \nabla_e \log q(\alpha;\ \mathbf{e})(\log p(\alpha,\ \mathcal{Y}_k) - \log q(\alpha;\ \mathbf{e})) \tag{6.46}$$

As a result, computing $\nabla_{\mathbf{e}}\mathcal{L}$ becomes a process of looking for an expectation, which can be estimated by the Monte Carlo approach. Then, the gradient descent method is used to make the parameters converge, which makes $q(\boldsymbol{\alpha};\mathbf{e})$ approximate the real posterior distribution.

The aforementioned black-box variational inference (BBVI) only relies on samples from $q(\boldsymbol{\alpha};\mathbf{e})$ to calculate the Monte Carlo gradient. However, the samples of the variational distribution $q(\boldsymbol{\alpha};\mathbf{e})$ are concentrated near the peak point, and there are fewer samples in the tail. This causes the Monte Carlo gradient to often have high variance, which means that the estimated gradient may be very different from the true value, resulting in optimization times that are too long.

In view of this defect, O-BBVI adds a good proposal distribution that matches the variational problem. $r(\boldsymbol{\alpha};\mathbf{e},\boldsymbol{\tau})$ is an overdispersed version of $q(\boldsymbol{\alpha};\mathbf{e})$. Its tail is heavier, which increases the probability that the value of the tail is sampled. In addition, O-BBVI also uses two strategies, control variates and Rao–Blackwellization in order to reduce the variance of the Monte Carlo gradient of the original BBVI. Finally, O-BBVI constructed a new Monte Carlo gradient equation, and based on the unconventional usage of importance sampling, it samples from $r(\boldsymbol{\alpha};\mathbf{e},\boldsymbol{\tau})$ and $q(\boldsymbol{\alpha};\mathbf{e})$, respectively, in order to estimate the gradient.

Let us begin with this overdispersed distribution

$$r(\boldsymbol{\alpha};\mathbf{e},\boldsymbol{\tau}) = g(\boldsymbol{\alpha},\boldsymbol{\tau})\exp\left\{\frac{\mathbf{e}^T t(\boldsymbol{\alpha}) - A(\mathbf{e})}{\boldsymbol{\tau}}\right\} \tag{6.47}$$

where $\boldsymbol{\tau} = [\tau_1, \tau_2]$ is the dispersion coefficient of the overdispersed distribution. For a fixed $\boldsymbol{\tau}$, $r(\boldsymbol{\alpha};\mathbf{e},\boldsymbol{\tau})$ and $q(\boldsymbol{\alpha};\mathbf{e})$ belong to the same exponential family. Moreover, $r(\boldsymbol{\alpha};\mathbf{e},\boldsymbol{\tau})$ allocates higher mass to the tails of $q(\boldsymbol{\alpha};\mathbf{e})$.

O-BBVI believes that for each component of the gradient, the proposal distribution that minimizes the variance of the estimator is not $q(\boldsymbol{\alpha};\mathbf{e})$ but the following.

$$q_n^{op}(\boldsymbol{\alpha}) \propto q(\boldsymbol{\alpha};\mathbf{e})|f_n(\boldsymbol{\alpha})| \tag{6.48}$$

That is, the optimal proposal distribution pushes the probability density to the tail of $q(\boldsymbol{\alpha};\mathbf{e})$;hence, we have reason to believe that $r(\boldsymbol{\alpha};\mathbf{e},\boldsymbol{\tau})$ is closer to the optimal proposal distribution than the variational distribution $q(\boldsymbol{\alpha};\mathbf{e})$. Essentially, $r(\boldsymbol{\alpha};\mathbf{e},\boldsymbol{\tau})$ provides a higher probability to be sampled for those $\boldsymbol{\alpha}$ that exist in the tail of the variational distribution, but have higher posterior probability. Thus, (6.45) can be rewritten as follows:

$$\hat{\nabla}_{\mathbf{e}}\mathcal{L} = \frac{1}{M}\sum_m f\left(\boldsymbol{\alpha}^{(m)}\right)\frac{q\left(\boldsymbol{\alpha}^{(m)};\mathbf{e}\right)}{r\left(\boldsymbol{\alpha}^{(m)};\mathbf{e},\boldsymbol{\tau}\right)} \tag{6.49}$$

where $\boldsymbol{\alpha}^{(m)}$ represents the m-th sample in the sample set of $r(\boldsymbol{\alpha};\mathbf{e},\boldsymbol{\tau})$.

When calculating the above equation using importance sampling, it is easy to fail due to the existence of high-dimensional hidden variables, but the related theory of mean field can be used to deal with this problem, i.e., keep the rest of the components unchanged when solving the n-th component. That is to say, (6.49) can be expressed as follows:

$$\hat{\nabla}_{e_n}\mathcal{L} = \mathbb{E}_{r(\alpha_n;e_n,\tau_n)}\left[w(\alpha_n)\mathbb{E}_{q(\boldsymbol{\alpha}_{-n};\mathbf{e}_{-n})}[f_n(\boldsymbol{\alpha})]\right] \tag{6.50}$$

where $w(\alpha_n) = q(\alpha_n; e_n)/r(\alpha_n; e_n, \tau_n)$, and $\boldsymbol{\alpha}_{-n}$ means variables other than α_n. $\mathbf{e}_{-n}$ is the same. Inspired by the foregoing discussion, a simple sample $\boldsymbol{\alpha}^0_{-n}$ is extracted from $q(\boldsymbol{\alpha}_{-n}; \mathbf{e}_{-n})$ to estimate $\mathbb{E}_{q(\boldsymbol{\alpha}_{-n};\mathbf{e}_{-n})}[\cdot]$ and M samples are extracted from $r(\alpha_n; e_n, \tau_n)$ to estimate $\mathbb{E}_{r(\alpha_n;e_n,\tau_n)}[\cdot]$. At the same time, a score function is defined as follows.

$$h_n\left(\alpha_n^{(m)}\right) = \nabla_{e_n} \log q\left(\alpha_n^{(m)}; e_n\right) \tag{6.51}$$

It is used to calculate the following.

$$f_n\left(\boldsymbol{\alpha}^{(m)}\right) = h_n\left(\alpha_n^{(m)}\right) \log \frac{p_n\left(\mathbf{y}_k, \alpha_n^{(m)}, \boldsymbol{\alpha}_{-n}^{(0)}\right)}{q\left(\alpha_n^{(m)}; e_n\right)} \tag{6.52}$$

This formula is the embodiment of Rao–Blackwellization. Note that Eq. (6.52) uses $\mathbf{y}_k$ instead of $\mathcal{Y}_k$ in order to reduce the complexity of the algorithm. Moreover, p_n is acquired by the Markov assumption. Herein, assume $\mathbf{x}_k$ is a Gaussian distribution with $\hat{\mathbf{x}}_k^{\alpha}$ as the mean and $\boldsymbol{P}_k^{\mathbf{x},\alpha}$ as the covariance. Given the state $\mathbf{x}_k$, the corresponding observation $\mathbf{y}_k$ obeys the Gaussian distribution $\mathcal{N}(\mathbf{y}_k;\boldsymbol{H}_k\mathbf{x}_k, \boldsymbol{R}^{\alpha_2})$. Thus, the likelihood function of the unknown parameter $\boldsymbol{\alpha}$ can be approximated as follows.

$$f(\mathbf{y}_k|\boldsymbol{\alpha}) = \mathcal{N}(\mathbf{y}_k; \boldsymbol{H}_k\mathbf{x}_k, \boldsymbol{R}^{\alpha_2}) \tag{6.53}$$

Consequently,

$$p_n\left(\mathbf{y}_k, \alpha_n, \boldsymbol{\alpha}_{-n}^{(0)}\right) = f(\mathbf{y}_k|\boldsymbol{\alpha})p_n(\alpha_n)p_{-n}\left(\boldsymbol{\alpha}_{-n}^{(0)}\right) \tag{6.54}$$

Finally, the gradient of ELBO relative to each component e_n is as follows:

$$\hat{\nabla}_{e_n}\mathcal{L}\ominus = \frac{1}{M}\sum_m \left[f_n^w\left(\boldsymbol{\alpha}^{(m)}\right) - b_n h_n^w\left(\alpha_n^{(m)}\right)\right] \tag{6.55}$$

where the following is the case.

$$f_n^w\left(\boldsymbol{\alpha}^{(m)}\right) = w\left(\alpha_n^{(m)}\right) f_n\left(\boldsymbol{\alpha}^{(m)}\right) \tag{6.56}$$

$$h_n^w\left(\alpha_n^{(m)}\right) = w\left(\alpha_n^{(m)}\right)h_n\left(\alpha_n^{(m)}\right) \tag{6.57}$$

$$b_n = \frac{Cov\left(f_n^w, h_n^w\right)}{Var\left(h_n^w\right)} \tag{6.58}$$

Note that b_n marks the application of the strategy of control variates.

Next, use the AdaGrad algorithm to make $\mathbf{e}$ converge to the optimum:

$$\mathbf{e}^{(t)} = \mathbf{e}^{(t-1)} + \boldsymbol{\lambda}_t \circ \hat{\nabla}_{\mathbf{e}}\mathcal{L} \tag{6.59}$$

where $\boldsymbol{\lambda}_t$ is the learning rate, '∘' stands for the Hadamard product. Note that the expectations contained in natural parameter $\mathbf{e}$ are the posterior noise statistics $\mathbb{E}_\alpha\left[\boldsymbol{Q}^{\alpha_1} | \mathbf{y}_k \right]$ and $\mathbb{E}_\alpha\left[\boldsymbol{R}^{\alpha_2} | \mathbf{y}_k \right]$ of the k-th sampling period and not the $\mathbb{E}_\alpha\left[\boldsymbol{Q}^{\alpha_1} | \mathcal{Y}_k \right]$ and $\mathbb{E}_\alpha\left[\boldsymbol{R}^{\alpha_2} | \mathcal{Y}_k \right]$. Here, they are only needed for computing the average value from 0 to the k-th sampling period.

Finally, if the posterior noise statistics converge to a certain value or changes little after multiple iterations, the subsequent state estimation can skip the step of finding the posterior noise parameters in order to save the algorithm overhead.

6.2.2 Experimental Comparison

This subsection will compare and analyze the performance differences between KFPNS and other Kalman filtering methods. The first Kalman filtering method to participate in the comparison is the IBR approach. The second one is the model-specific Kalman filtering approach, which is a classical Kalman filter designed with respect to real noise parameters. The third is the minimax method, which performs best in the worst case when the noise parameters are uncertain. The parameters of the minimax method are set to $\alpha_{\max 1} = 4$, $\alpha_{\max 2} = 4$. The last one is OBKF, which designs a message passing algorithm based on factor graphs to calculate the likelihood function formulaically and then employs the Metropolis Hastings MCMC method to find posterior effective noise statistics.

1. Simulation

For the tracking scenario in a two-dimensional space, the state of the target is $\mathbf{x}_k = \left[p_x p_y v_x v_y\right]^T$, and the subscripts x, y represent the x and y dimensions, respectively. Assuming that the target's motion model is a constant velocity model and that there is only a weak random disturbance, the radar obtains the measurements of the target at every t seconds. Based on (9) and (10), the state-space model of the measured target is constructed, and the matrices are as follows.

$$\boldsymbol{\Phi}_k = \begin{bmatrix} 1 & 0 & t & 0 \\ 0 & 1 & 0 & t \\ 0 & 0 & 1 & 0 \\ 0 & 0 & 0 & 1 \end{bmatrix}, \ \boldsymbol{H}_k = \begin{bmatrix} 1 & 0 & 0 & 0 \\ 0 & 1 & 0 & 0 \end{bmatrix}, \ \boldsymbol{\Gamma}_k = \begin{bmatrix} t^2/2 & 0 \\ 0 & t^2/2 \\ t & 0 \\ 0 & t \end{bmatrix} \tag{6.60}$$

The covariance matrices of the process and observation noise are expressed as follows.

$$\boldsymbol{Q}^{\alpha_1} = \begin{bmatrix} \alpha_1 & 0 \\ 0 & \alpha_1 \end{bmatrix}, \ \boldsymbol{R}^{\alpha_2} = \begin{bmatrix} \alpha_2 & 0 \\ 0 & \alpha_2 \end{bmatrix} \tag{6.61}$$

Let $t = 1$ and the target's motion state is initialized to $\mathbb{E}[\mathbf{x}_0] = [1035 - 5]^T$ and $\mathbb{E}_\alpha\left[\boldsymbol{P}_0^{\mathbf{x},\boldsymbol{\alpha}}|\mathcal{Y}_{-1}\right] = \mathrm{diag}([10001000100100])$, where $\mathrm{diag}(\mathbf{a})$ represents a diagonal matrix with elements in vector **a** as diagonal elements.

In the first simulation, it is assumed that both α_1 and α_2 are unknown parameters that are uniformly distributed in the interval [1,4] and [0.5,4], respectively. The generating a certain amount of data based on the above model is the necessary condition for analyzing the average performance of the five Kalman filtering methods, and the evaluation indicators are their respective average MSE. The results are shown in Fig. 6.13. This set of simulation data is generated according to the prior distribution of noise, that is, randomly generates 200 pairs of $[\alpha_1,\alpha_2]$, and each combination corresponds to 10 different observation sequences. Among the five filters, the average MSE of the model-specific Kalman filter is the smallest because it uses real noise parameters. Instead, the minimax approach only considers the worst case and cannot effectively solve the problem of unknown noise second-order statistics. Figure 6.13 also suggests that the average MSEs of OBKF and IBRKF are almost the same at the initial stage of filtering, and both greatly outperform KFPNS since the performance of KFPNS depends on the length of the observation sequence used to adjust filter parameters and the prior distribution of noise. If the number of sampling period k is small, the filter parameters may be adjusted incorrectly after each iteration, resulting in large state estimation errors for certain subsequent sampling periods. However, in the long run, the increase in analyzable data will return the uncertain parameters close to the true value. At this time, the performance of the KFPNS is even similar to that of OBKF. Both KFPNS and OBKF need to use a certain amount of observed data to estimate unknown noise parameters in order to achieve better estimation results, which causes them to converge more slowly than other algorithms, especially for KFPNS. The good news is that as the number of observations continues to increase, their performance is closer to the model-specific Kalman filter.

The second simulation sets two specific noise parameter combinations. Each specific combination provides 200 observation sequences to weaken the interference of abnormal results. Then, calculate the average MSE to directly display the performance differences of each filter. The first column of Fig. 6.14 is a performance analysis based on the specific model $\alpha_1 = 3, \ \alpha_2 = 1.5$. The true value of the specific noise pair in the second column is $\alpha_1 = 3.5, \ \alpha_2 = 3.5$. The variation of average

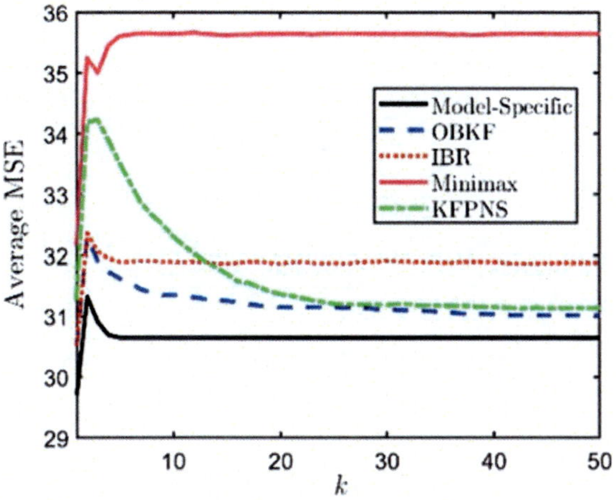

Fig. 6.13 Average MSE of various filters when α_1 and α_2 are uncertain

MSE of each filter in Fig. 6.14a is roughly consistent with the previous simulation. Subgraph (c) and (e), respectively, show the variation of the average posterior mean $\mathbb{E}[\alpha_1|\mathcal{Y}_k]$ and $\mathbb{E}[\alpha_2|\mathcal{Y}_k]$ under the first specific model. Similarly, (d) and (f) correspond to the second specific noise combination. The vertical columnar line represents the variance, and the size of variance is represented by the length of the line. The figures in the second and third rows show that with the increase in observation data, $\mathbb{E}[\alpha_1|\mathcal{Y}_k]$ and $\mathbb{E}[\alpha_2|\mathcal{Y}_k]$, moves closer to the true value from the preset empirical values of 2.2 and 2 and finally stabilizes near the true value. At the same time, the variance of the average posterior also means changes from small to large and then to small and eventually stabilizes. This indicates that the posterior distribution of noise estimated by KFPNS is more and more in line with the true distribution, and its value is concentrated around the true value. It can also be observed from Subgraph (b) that the minimax method outperforms IBRKF and even outperforms OBKF in the beginning. Note that under the condition of prior knowledge, the average performance of IBRKF exceeds the minimax Kalman filter. However, IBRKF performs worse with respect to some specific models, for example, when the parameters of the minimax method are closer to the true values than the prior mean of IBRKF. Fortunately, both the OBKF and KFPNS use their approaches to approximate true posterior noise distribution, and with the input of more data, posterior second-order statistics will tend towards the true model. This allows them to break the limitations of most noise models and to deal with models that IBRKF is unable to handle.

When the mass of the prior probability is not uniform, the performance changes of different robust Kalman filters are also worth discussing. Suppose α_1 is uniformly distributed in the interval $[1, 2]$ and fixed. On the other hand, let a Beta distribution $\mathcal{B}(\alpha_r, \beta_r)$ in the interval $[0.25, 4]$ govern α_2. The mean and variance of the Beta distribution are $\alpha_r / \alpha_r + \beta_r$ and $\alpha_r\beta_r / (\alpha_r + \beta_r)^2(\alpha_r + \beta_r + 1)$, respectively, and $\alpha_r + \beta_r = 1$. On the basis of ensuring that the mean of this prior distribution remains

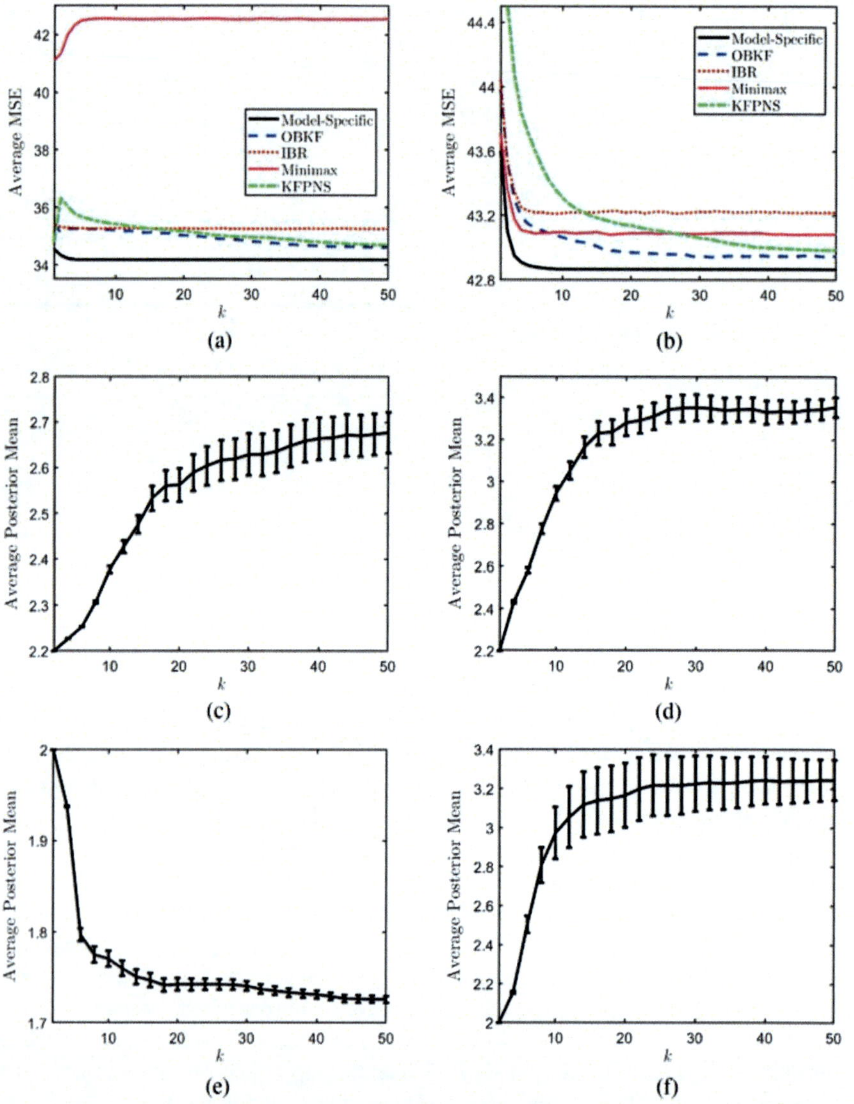

Fig. 6.14 Performance analysis for specific noise pairs. **a** the average MSE for specific noise model $\alpha_1 = 3,\ \alpha_2 = 1.5$. **b** the average MSE for specific noise model $\alpha_1 = 3.5,\ \alpha_2 = 3.5$. **c** the variation of $\mathbb{E}[\alpha_1|\mathcal{Y}_k]$ and its variance when $\alpha_1 = 3$. **d** the variation of $\mathbb{E}[\alpha_1|\mathcal{Y}_k]$ and its variance when $\alpha_1 = 3.5$. **e** the variation of $\mathbb{E}[\alpha_2|\mathcal{Y}_k]$ and its variance when $\alpha_2 = 1.5$. **f** the variation of $\mathbb{E}[\alpha_2|\mathcal{Y}_k]$ and its variance when $\alpha_2 = 3.5$

unchanged, a new prior distribution with different mass can be obtained by appropriately adjusting its parameters. Consider two pairs of specific parameters in this simulation, $\alpha_r = 0.1,\ \beta_r = 0.9$ and $\alpha_r' = 0.1\ \alpha_r,\ \beta_r' = 0.1\beta_r$. As a result, the reduction in parameters causes the variance to become larger, and the prior distribution will be more relaxed. Figure 6.15 shows their average MSEs. Even if the Beta distribution is changed to make the probability of values in prior distribution different, the average MSEs of the minimax method and IBRKF are almost unchanged. The reason is that for all prior distributions, the prior mean of the IBR method is always the same, while the minimax method does not consider prior distribution at all. They cannot effectively cope with sudden changes in the model. In this case, they are equivalent to the classical Kalman filter. Compared with the above two robust filtering strategies, although the mass distribution of the prior has changed, since KFPNS incorporates observations into prior knowledge in order to estimate posterior noise distribution, its average MSE is related to the tightness of the prior distribution. In other words, the tighter the prior distribution, the worse the performance of KFPNS, while the opposite is true when the prior distribution is relaxed. Importantly, the average MSE of KFPNS can always be closer to that of the model-specific Kalman filter.

The last simulation discusses the performance of each robust filtering strategy when the prior knowledge is inaccurate. In practical applications, it is often impossible to understand the underlying true model, and the mastery of prior knowledge is not comprehensive. Therefore, it is crucial for a robust filter not to rely too much on the prior distribution. The previous simulation is based on the assumption that the prior distribution of noise is accurate and available. Here, only the case where the variance of the observation noise is unknown and the range of its prior distribution is included in the interval of the true distribution is considered. It is still assumed that α_1 uniformly distributed over $[1, 2]$, while α_2 is uniformly distributed over $[3, 5]$, which is wrong relative to the real interval. Similarly, 20 pairs of different true values are generated according to the correct prior interval, and each pair of true values contains 20 sets of different observations. The evaluation index of the robustness of each filter

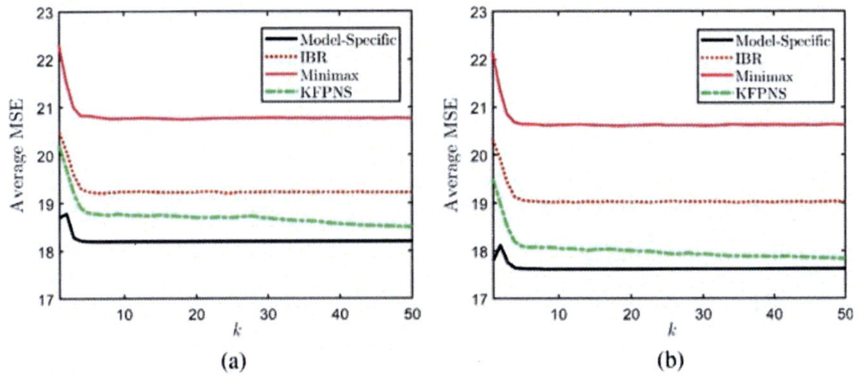

Fig. 6.15 The average performance of four filters with different Beta priors. **a** $\alpha_r = 0.1,\ \beta_r = 0.9$; **b** $\alpha_r' = 0.1\alpha_r,\ \beta_r' = 0.1\beta_r$

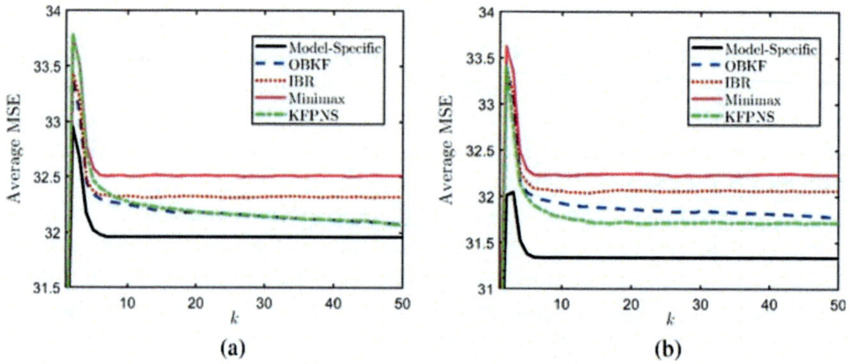

Fig. 6.16 Average MSE of various filters when α_2 is uncertain, and the interval $[3, 5]$ of prior distribution is inaccurate. **a** the exact interval is $[2, 6]$. **b** the exact interval is $[0.5, 7.5]$

is still the average MSE, and the results are shown in Fig. 6.16. Subgraph (a) and (b) are the average MSEs of various Kalman filters in the entire sampling period when the correct interval is $[2, 6]$ and $[0.5, 7.5]$. When k is large, the two robust filters, OBKF and KFPNS still perform best in terms of average MSEs. It is worth noting that KFPNS outperforms OBKF for the first time in subgraph (b), but it is slightly inferior or equivalent to OBKF in the previous simulation. This is related to the computing method of posterior noise statistics. When OBKF employs the Metropolis Hastings MCMC method to select sample points, the prior distribution incorporated is wrong and lacks an adjustment strategy, which makes the posterior noise statistics unable to converge to the true value. On the contrary, KFPNS adds a weighting factor in the calculation process, making it not completely dependent on the inaccurate prior distribution. Therefore, KFPNS performs better when facing models with larger errors between prior knowledge and the underlying true model. Unfortunately, the performance of KFPNS may decline when the prior distribution provided is inaccurate, and the stronger the inaccuracy, the faster the performance decline.

2. Experiment

The observed data of a real MIMO radar system are used to verify the performance of the proposed method. The experimental conditions are described as follows. First, delimit two fixed experimental areas, namely the indoor area of 175×465 cm^2 and the outdoor area of $800 \times 800 cm^2$. Secondly, some positions in the two experimental areas are calibrated, and these positions are regarded as the ground truth of the target. Table 6.7 reports the parameters of the radar used. Next, the environment within the experimental area is sampled without any targets, and non-target points are filtered out based on this. During data acquisition, two people moved at a constant velocity in these two areas according to two predetermined trajectories. Figure 6.17 depicts the above two experimental scenarios. After the measurements of two targets are collected, these measurements are clustered, and the cluster centers are used as

Table 6.7 MIMO radar parameters

Radar model	Radar system	The start frequency	Range resolution	Frame periodicity	Scan range
RDP 77S244-ABM-AIP	FMCW	77 GHz	0.045 m	200 ms	FOV 120°

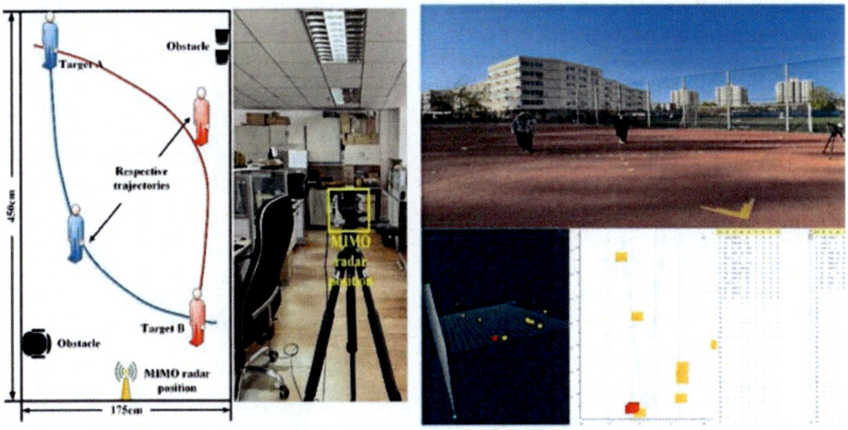

(a) (b)

Fig. 6.17 Real data acquisition scenes. **a** indoor scene. **b** outdoor scene

the observation positions of the two targets. Finally, four different Kalman filtering methods are used for data processing and error analysis.

The filtering results under different experimental scenarios are shown in Figs. 6.18 and 6.19. Note that the blue point represents the estimated position of target A, and the red point represents the estimated position of target B. When filtering these position data, we do not fully grasp the underlying real model, which means that the prior knowledge we provided for these algorithms may not be accurate enough. Nevertheless, it can be observed that the estimated trajectories of the two Kalman filters designed based on posterior information are significantly smoother than the other methods. The inaccuracy of prior knowledge of noise is the main reason for the poor trajectory estimation of CKF and IBRKF. Although IBRKF introduces the concept of effective statistics, which weakens this effect to a certain extent, when the inaccuracy of prior knowledge is strong, its performance is only slightly improved compared with the classical algorithm. As expected, both KFPNS and OBKF have the characteristics of computing the posterior value and making it approximate the real value through algorithm iteration, which makes them very robust; thus, their estimated trajectory is also the closest to the true trajectory.

In addition, the position error (PE) between the estimated position and the true position will also be used as an evaluation index for filter performance. In Fig. 6.20, although the PE of the new filter sometimes exceeds IBRKF and the classical Kalman

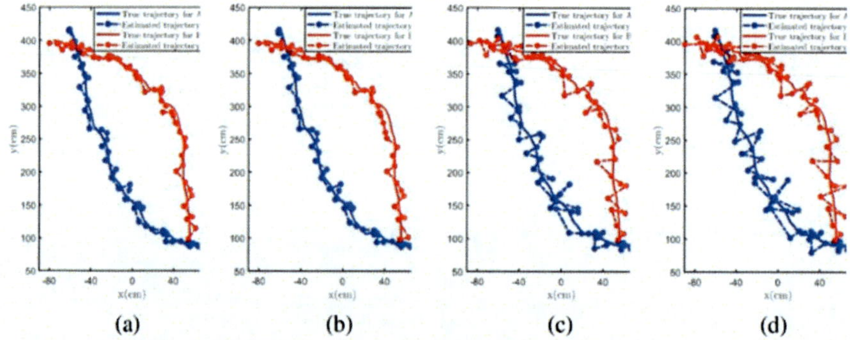

Fig. 6.18 The fitting degree between the estimated trajectory and the true trajectory after processing by various Kalman filtering methods in an indoor scene. **a** OBKF. **b** KFPNS. **c** IBRKF. **d** CKF

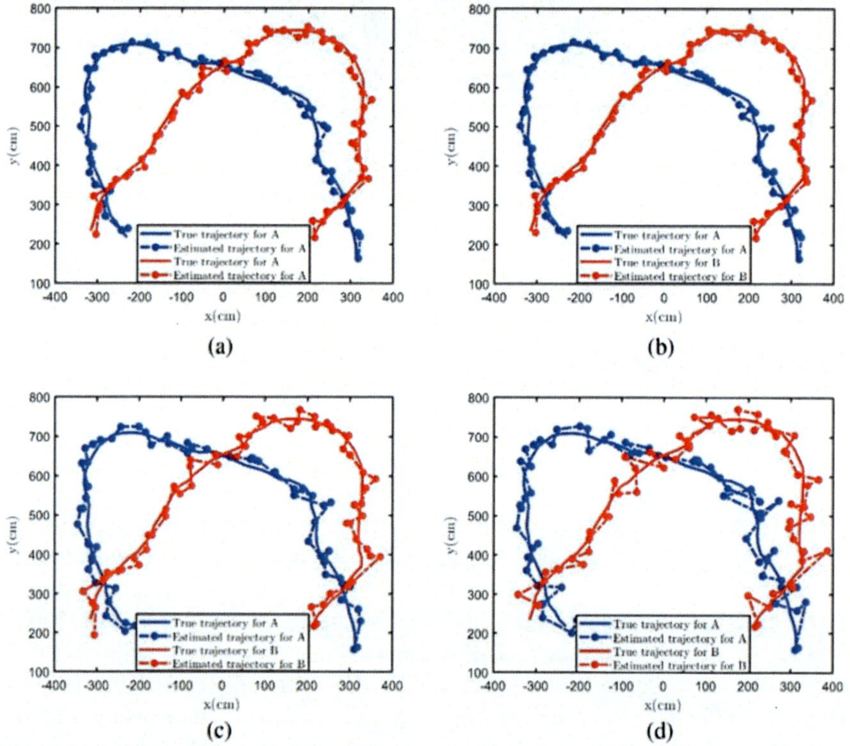

Fig. 6.19 The fitting degree between the estimated trajectory and the true trajectory after processing by various Kalman filtering methods in an outdoor scene. **a** OBKF. **b** KFPNS. **c** IBRKF. **d** CKF

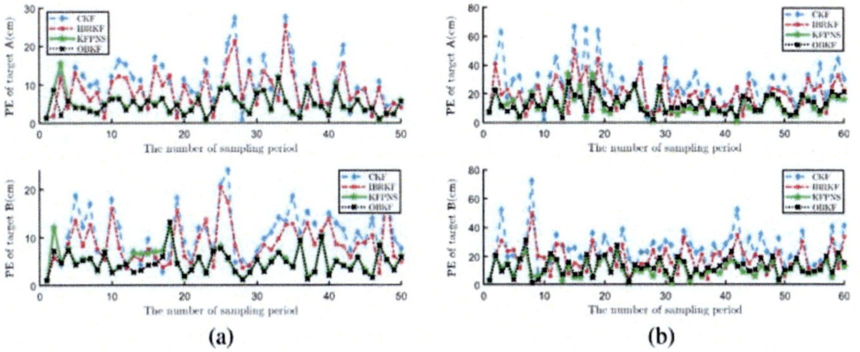

Fig. 6.20 The PE of each filter in two specific scenes. **a** comparison of PE of various Kalman filters in an indoor scene. **b** comparison of PE of various Kalman filters in an outdoor scene

filter, overall, most of the PE curve of KFPNS is in a lower position. Moreover, the PE of the proposed algorithm is more concentrated than compared to these two algorithms. However, in subgraph (a), it cannot be ignored that KFPNS performs poorly in the early stages of filtering. The reason for this phenomenon may be that the selection of the initial value is quite different from the real value, and there are fewer observed data available for analysis. Fortunately, with the input of more observed data, its estimation error gradually approaches OBKF.

The root-mean-square error (RMSE) of the four algorithms are shown in Fig. 6.21. In Subgraph (a), when the number of sampling period is small, the RMSE of KFPNS takes the maximum value. After 20 sampling periods, the RMSE of KFPNS tends to be stable and is significantly different from that of the other two Kalman filters with poor robustness. Subgraph (b) also suggests that the gap between KFPNS and OBKF is further narrowed. The change of RMSE over time proves that both KFPNS and OBKF have strong robustness. Table 6.8 reports the mean value of RMSE (MMSE) for various algorithms, which can be used to intuitively compare the average performance of various filters in different scenes. In the indoor scene, for target B, the average performance of KFPNS is 29.62% and 41.56% higher than that of IBRKF and the classical Kalman filter separately. Despite the average performance of KFPNS decreases by 18.1% compared with OBKF, they differ by only 0.9434 cm.

Figure 6.22 further analyzes the cumulative distribution function (CDF) of the RMSE of each filtering algorithm in the indoor scene. In Fig. 6.22, 90% of the RMSE of KFPNS is less than 6.739 cm, which improves by 31% over IBRKF (9.767 cm) and 42.35% over CKF (11.69 cm) but degrades by 23.11% over OBKF (5.474 cm). In a nutshell, compared with the IBR robust filtering strategy, KNPNS and OBKF are more suitable for models where the noise is unknown and the real prior cannot be fully grasped.

3. Time cost analysis

Previous experiments have proved that OBKF and KFPNS have considerable robustness, so their complexity needs to be further compared. Since the recursive structure

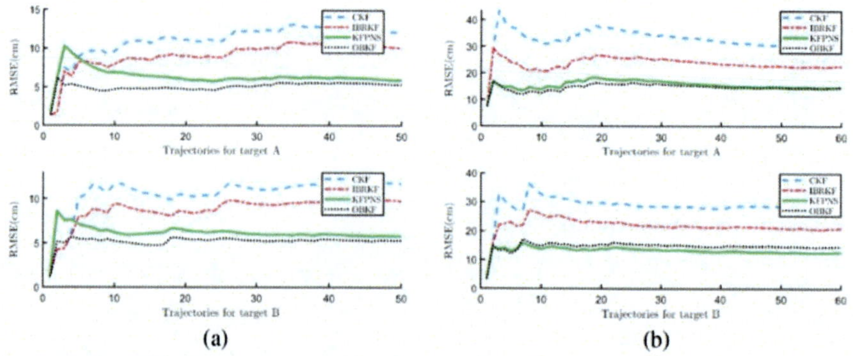

Fig. 6.21 The RMSE of various Kalman filtering methods in different scenes. **a** the RMSE of the four Kalman filtering methods in an indoor scene with respect to target A and target B, respectively. **b** the RMSE of the four Kalman filtering methods in an outdoor scene with respect to target A and target B, respectively

Table 6.8 The MMSE difference of various Kalman filtering methods

Algorithm	MMSE			
	Indoor scene		Outdoor scene	
	Target A (cm)	Target B (cm)	Target A (cm)	Target B (cm)
KFPNS	6.3679	6.1552	15.5520	13.1918
IBRKF	9.0658	8.7459	23.5724	21.7488
CKF	11.0092	10.5480	32.6647	28.4253
OBKF	5.0358	5.2118	14.6269	14.6148

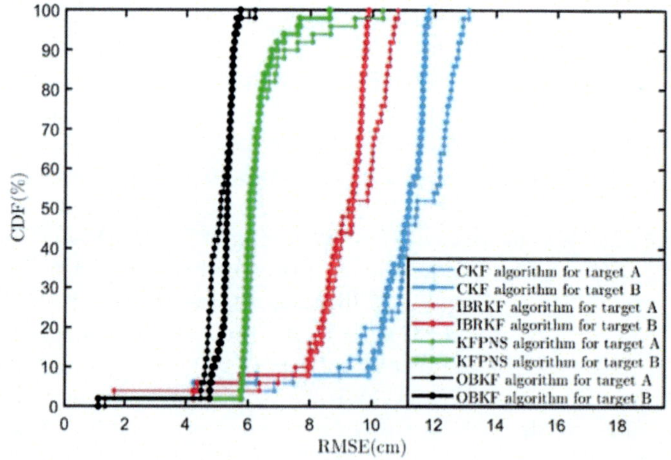

Fig. 6.22 CDFs of RMSE for various algorithms in an indoor scene

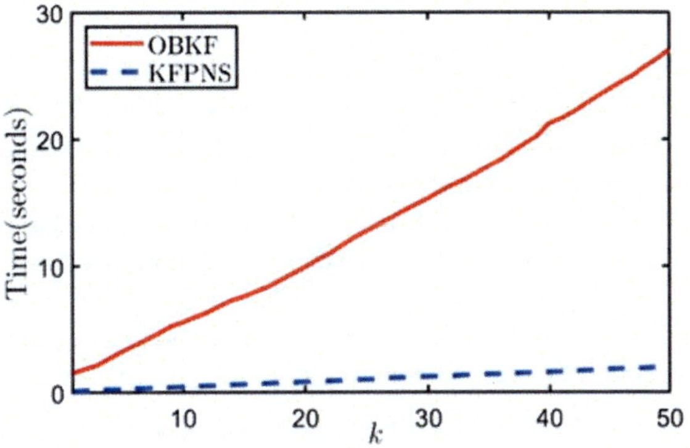

Fig. 6.23 The time consumption of two algorithms when the length of the observation sequence changes

of these two robust Kalman filters is consistent with the classical algorithm, their computational burden is mainly concentrated on the calculation of posterior noise statistics. It is worth mentioning that OBKF uses a factor-graph-based method to convert the problem of finding the likelihood function into a matrix operation, which increases the complexity of the algorithm. Moreover, if the posterior noise statistics at the k-th observation need to be computed, the likelihood function of each MCMC sample must be iterated from $i = 0$ until $i = k - 1$. Therefore, its computational complexity depends on the dimension of the matrix, the number of samples in the posterior distribution and the number of sampling periods. In contrast, the proposed algorithm rarely involves matrix operations, and there is no need to compute the likelihood function from the beginning. At the same time, the variational distribution is fixedly decomposed into two factors. These reasons all reduce the calculation burden of KFPNS to a considerable extent. Figure 6.23 shows the average run times of the OBKF and KFPNS relative to the sampling period. Note that both algorithms need to generate 10,000 samples to approximate the true posterior noise distribution. Figure 6.23 illustrates that under the same number of samples and the same length of observation sequence, OBKF needs to consume more time. Consequently, KFPNS performs better in real time and is more suitable for applications in actual projects.

6.3 Data Association Algorithms

6.3.1 Nearest Neighbor Data Association

The function of the data association algorithm is to associate the observation point at the current moment with the prediction point. Sample points that are successfully associated will continue to be used, and sample points that fail to be associated will no longer be used. The data association and filtering prediction process between the sample points and the track are carried out simultaneously.

The Nearest Neighbor Data Association (NNDA) algorithm is simple and fast, which takes the predicted point coordinates as the center of the correlation gate to automatically generate the correlation gate. The correlation gate must contain the real observation points of the target. Based on the gate, the points not in the gate are judged as clutter or false alarm signals, and these points are not related to the track. Kalman filter is used to predict the target position. At this time, the normalized statistical distance d_{ij}^2 between the observation point and the prediction point is:

$$d_{ij}^2 = \boldsymbol{e}_{ij}(k)\boldsymbol{S}_{ij}^{-1}(k)\boldsymbol{e}_{ij}^{\mathrm{T}}(k) \tag{6.62}$$

where, i represents the i-th target, j represents the j-th observation point, $\boldsymbol{e}_{ij}(k)$ is the residual, that is, the observation error matrix, and the covariance matrix of $\boldsymbol{e}_{ij}(k)$ is $\boldsymbol{S}_{ij}(k)$. The residual at time k is the difference between the observed value $\mathbf{Z}_j(k)$ at time k and the predicted value $\hat{\mathbf{Z}}_i(k|k-1)$.

If the observation point falls within the correlation gate, it should satisfy:

$$d_{ij}^2 \le \gamma \tag{6.63}$$

where, γ is determined according to the actual situation.

Expand Eq. (6.22) to get:

$$d_{ij}^2 = [\mathbf{Z}_j(k) - \hat{\mathbf{Z}}_i(k|k-1)]\boldsymbol{S}_{ij}^{-1}(k)[\mathbf{Z}_j(k) - \hat{\mathbf{Z}}_i(k|k-1)]^{\mathrm{T}} \tag{6.64}$$

There are two main situations in the correlation process: ① If there is only one observation point in the correlation gate, the observation point is considered to be the real observation point, and the observation point is directly used for the next operation; ② If there is more than one observation point in the correlation gate, the statistical distance between all the observation points and the predicted points that fall within the correlation gate should be calculated, and the observation point with the smallest statistical distance should be selected as the valid point, whose valid information will be used to update the track. The nearest neighbor data association is shown in Fig. 6.24.

6.3.2 *Joint Probabilistic Data Association*

The Joint Probabilistic Data Association (JPDA) algorithm based on the idea of probability was proposed by Bar-Shalom and developed from the PDA algorithm. The PDA algorithm has higher accuracy in single-target data association, but it does not take into account that the observation points are in the multi-target correlation gate intersection area, so in the multi-target situation, the PDA algorithm cannot accurately associate. However, considering all situations, by combining all the events from the clutter or the target to form a joint event, the JPDA calculates the probability

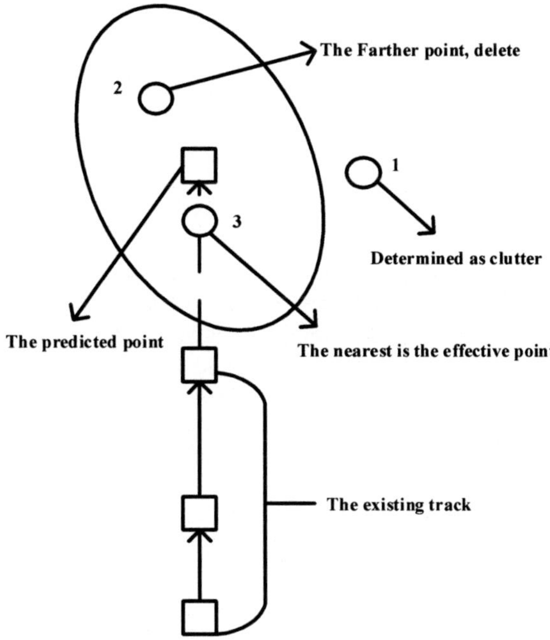

Fig. 6.24 The association process of the nearest neighbor data association

of the joint event and selects the event with the highest probability to correct the track. Therefore, the JPDA is suitable for multi-target tracking in clutter, which solves the problem that the PDA algorithm is easy to lose or follow.

The steps of the JPDA are as follows.

Build a validation matrix

$$\boldsymbol{\Omega} = [\omega_{jt}] = \begin{bmatrix} \omega_{10} & \omega_{11} & \omega_{12} & \cdots & \omega_{1T} \\ \omega_{20} & \omega_{21} & \omega_{22} & \cdots & \omega_{2T} \\ \vdots & \vdots & \vdots & & \vdots \\ \omega_{m_k0} & \omega_{m_k1} & \omega_{m_k2} & \cdots & \omega_{m_kT} \end{bmatrix} \tag{6.65}$$

where, j represents the observation point ($j = 1, 2, \ldots, m_k$), t represents the target ($t = 0, 1, \ldots, T$), $\omega_{jt} = 1$ represents the observation point j falls within the associated gate of the target t, and $\omega_{jt} = 0$ is the opposite; m_k is the number of valid observation points, T is the number of targets, and the first column of $\boldsymbol{\Omega}$ indicates that each observation point may originate from background clutter, and all elements in the first column are 1.

Let $\theta(k)$ be the set of all joint events at time k, and use n_k to represent the number of joint events, then we have

$$\theta(k) = \{\theta_i(k)\}_{i=1}^{n_k} \tag{6.66}$$

Split the formula (6.25) to get the connected matrices

$$\hat{\boldsymbol{\Omega}}[\theta_i(k)] = \left(\hat{\omega}^i_{jt}\right) = \begin{bmatrix} \hat{\omega}^i_{10} & \hat{\omega}^i_{11} & \hat{\omega}^i_{12} & \cdots & \hat{\omega}^i_{1T} \\ \hat{\omega}^i_{20} & \hat{\omega}^i_{21} & \hat{\omega}^i_{22} & \cdots & \hat{\omega}^i_{2T} \\ \vdots & \vdots & \vdots & & \vdots \\ \hat{\omega}^i_{m_k 0} & \hat{\omega}^i_{m_k 1} & \hat{\omega}^i_{m_k 2} & \cdots & \hat{\omega}^i_{m_k T} \end{bmatrix} \tag{6.67}$$

where $\hat{\omega}^i_{jt}$ is a binary element, which takes the value 1 when the observation point j originates from the target t, and 0 otherwise.

Assume that the event of observation point j originating from target t is $\theta_{jt}(k)$, and the event of observation point j originating from background clutter is $\theta_{0t}(k)$, then the probability $\beta_{jt}(k)$ of observation point and target interconnection is

$$\beta_{jt}(k) = P\left\{\theta_{jt}(k)|\mathbf{Z}^k\right\} \tag{6.68}$$

where, $\mathbf{Z}^k$ is all return waves that fall within the correlation gate from the start of the track to the moment. From formula (6.28), it can be known that

$$\sum_{j=0}^{m_k} \beta_{jt}(k) = 1 \tag{6.69}$$

Let the state estimate of the target t at time k be $\hat{\boldsymbol{X}}^t(k|k)$, and its expression is

$$\begin{aligned} \hat{\boldsymbol{X}}^t(k|k) &= E\left[\boldsymbol{X}^t(k)|\mathbf{Z}^k\right] \\ &= \sum_{j=0}^{m_k} E\left[\boldsymbol{X}^t(k)|\theta_{jt}(k), \mathbf{Z}^k\right] P\left\{\theta_{jt}(k)|\mathbf{Z}^k\right\} \\ &= \sum_{j=0}^{m_k} \hat{\boldsymbol{X}}^t_j(k|k)\beta_{jt}(k) \end{aligned} \tag{6.70}$$

The state of the $j-th$ observation point and target t obtained by filtering at time k is estimated as

$$\hat{\boldsymbol{X}}^t_j(k|k) = E\left[\boldsymbol{X}^t(k)|\theta_{jt}(k), \mathbf{Z}^k\right] \tag{6.71}$$

Therefore, the probability $\beta_{jt}(k)$ of interconnection is

$$\begin{aligned} \beta_{jt}(k) &= P\left\{\theta_{jt}(k)|\mathbf{Z}^k\right\} \\ &= P\left\{\bigcup_{i=1}^{n_k} \theta^i_{jt}(k)|\mathbf{Z}^k\right\} \\ &= \sum_{i=1}^{n_k} \hat{\omega}^i_{jt} P\left\{\theta_i(k)|\mathbf{Z}^k\right\} \end{aligned} \tag{6.72}$$

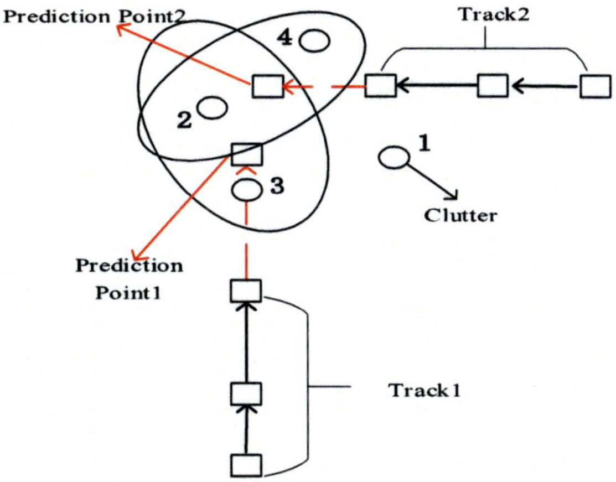

Fig. 6.25 The association method of the joint probability data association

The state estimation covariance at time k is

$$P_j^t(k|k) = E\left\{\left[\boldsymbol{X}^t(k) - \hat{\boldsymbol{X}}_j^t(k|k)\right]\left[\boldsymbol{X}^t(k) - \hat{\boldsymbol{X}}_j^t(k|k)\right]^{\mathrm{T}} |\theta_{jt}(k), \boldsymbol{Z}^k\right\} \tag{6.73}$$

The following formula is obtained by Kalman filter

$$P_j^t(k|k) = P^t(k|k-1) - \boldsymbol{K}^t(k)\boldsymbol{S}^t(k)\boldsymbol{K}^{t'}(k) \tag{6.74}$$

In the formula, $\boldsymbol{K}^t(k)$ is the gain matrix, and $\boldsymbol{S}^t(k)$ is the innovation covariance matrix. The association method of the joint probability data association is shown in Fig. 6.25.

6.3.3 K-Nearest Neighbor Joint Probabilistic Data Association Algorithm

In order to meet the requirements of traffic surveillance radar for vehicle tracking accuracy and operation efficiency, this section improves the multi-target tracking data association algorithm. Combined with the advantages of NNDA algorithm and JPDA algorithm, a k-nearest neighbor joint probability data association algorithm is proposed. The algorithm flow is as follows.

(1) Screen out the observation points associated with the prediction points of each target vehicle in the observation data, and the prediction point of each target vehicle at the current moment is determined according to the observation data at the previous moment. When the radar detects the target vehicle, in addition to receiving the echo signal of the target vehicle, it will also receive noise, clutter or interference signals. Therefore, in order to reduce the amount of calculation

and improve the vehicle tracking accuracy, before using the observation data to track the vehicle, it is necessary to judge whether the observation points in the observation data are valid.

Use the NNDA algorithm to select the center of the prediction point, create a correlation gate, and based on the preset threshold of each correlation gate, according to the distance between each observation point in the observation data at the current moment and the predicted point of the target vehicle, determine that the current moment falls into each correlation gate observation point inside. When selecting the correlation gate threshold, it is necessary to make the correlation gate contain the real observation points of the target vehicle, and the points not in the gate are judged as clutter or false alarm signals, and these points are not related to the track. The position of the target vehicle is predicted by using Kalman filter. At this time, the normalized statistical distance between the observation point and the predicted point is

$$d_{ji}^2 = \boldsymbol{e}_{ji}(t)\boldsymbol{S}_{ji}^{-1}(t)\boldsymbol{e}_{ji}^{\mathrm{T}}(t) \tag{6.75}$$

In the formula, d_{ji}^2 is the statistical distance from the j-th observation point to the predicted point of the i-th target vehicle; $\boldsymbol{e}_{ji}(t)$ is the residual, that is, the observation error matrix is the difference between the observed value $\mathbf{Z}_j(t)$ and the predicted value $\hat{\mathbf{Z}}_i(t|t-1)$ at time t; $\boldsymbol{S}_{ji}(t)$ is the covariance matrix of $\boldsymbol{e}_{ji}(t)$.

If the observation point falls within the correlation gate, it should satisfy:

$$d_{ji}^2 \le \gamma \tag{6.76}$$

where, it is determined according to the actual situation.

Expand Eq. (6.36) to get:

$$d_{ji}^2 = [\mathbf{Z}_j(t) - \hat{\mathbf{Z}}_i(t|t-1)]\boldsymbol{S}_{ji}^{-1}(t)[\mathbf{Z}_j(t) - \hat{\mathbf{Z}}_i(t|t-1)]^{\mathrm{T}} \tag{6.77}$$

Calculate the statistical distance d_{ji}^2 between all observation points and the predicted point of the current target vehicle. If $d_{ji}^2 \le \gamma$, mark it as a valid observation point and save all its information; if $d_{ji}^2 > \gamma$, determine it as clutter and filter it out directly.

(2) Construct a first pre-aggregation matrix according to the relationship between the selected observation points and the predicted points of the target vehicle. One row of the first pre-aggregation matrix represents an observation point in the observation data, and each column represents of the first pre-aggregation matrix a predicted point for a target vehicle.

Let the number of observation points be m_k, j represents the observation point ($j = 1,\ 2, \ldots, m_k$); the number of prediction points of the target vehicle is I, and i represents the prediction point of the target vehicle ($i = 0,\ 1, \ldots, I$), and $i = 0$ represents the 0-th target vehicle, that is $i = 0$, the background clutter. In order to realize the screening of observation points, a statistical distance weight is introduced here, and a first pre-aggregation matrix is constructed according to the relationship between the selected observation points and the predicted points.

(3) Calculate the statistical distance weight from each observation point to the predicted point, and sort the rows of the first pre-aggregation matrix in the order of the statistical distance weight from small to large to obtain the sorted second pre-aggregation matrix.

The weight of the statistical distance from each observation point to the predicted point is:

$$p(d_{ji}^2) = \frac{d_{ji}^2}{\sum\limits_{j=1}^{m_k} d_{ji}^2} \tag{6.78}$$

In the formula, $p(d_{ji}^2)$ represents the statistical distance weight from the $j-th$ observation point to the prediction point of the $i-th$ target vehicle; m_k represents the number of observation points; I represents the number of prediction points; d_{ji}^2 represents the j-th observation point to the i-th target Statistical distance of the predicted points of the vehicle. The minimum statistical distance d_{ji}^2 between the j-th observation point and the i-th target vehicle's predicted point is used as the calculation standard.

The first pre-aggregation matrix $\boldsymbol{\Delta}'$ is:

$$\boldsymbol{\Delta}' = \left[d_{ji}^2 | p(d_{ji}^2)\right] = \left[\begin{array}{cccc|c} \omega_{11} & \omega_{12} & \cdots & \omega_{1I} & p(d_{1i}^2) \\ \omega_{21} & \omega_{22} & \cdots & \omega_{2I} & p(d_{2i}^2) \\ \vdots & \vdots & & \vdots & \vdots \\ \omega_{m_k I} & \omega_{m_k I} & \cdots & \omega_{m_k I} & p(d_{m_k i}^2) \end{array}\right] \tag{6.79}$$

where, ω_{ji} represents the relationship between the j-th observation point and the i-th target vehicle prediction point. For each element in the first pre-aggregated matrix $\boldsymbol{\Delta}$, if $d_{ji}^2 \leq \gamma, \omega_{ji} = 1$; if $d_{ji}^2 > \gamma, \omega_{ji} = 0$.

The second pre-aggregation matrix $\boldsymbol{\Delta}'$:

$$\boldsymbol{\Delta}' = \left[d_{ji}^2 | p(d_{ji}^2)\right] = \left[\begin{array}{cccc|c} \omega_{11} & \omega_{12} & \cdots & \omega_{1I} & p(d_{1i}^2) \\ \vdots & \vdots & & \vdots & \vdots \\ \omega_{k1} & \omega_{k2} & \cdots & \omega_{kI} & p(d_{ki}^2) \\ \vdots & \vdots & & \vdots & \vdots \\ \omega_{m_k 1} & \omega_{m_k 2} & \cdots & \omega_{m_k I} & p(d_{m_k i}^2) \end{array}\right] \tag{6.80}$$

Here, it is assumed that the order of the statistical distance weights is $p(d_{1i}^2) \leq \ldots \leq p(d_{ki}^2) \leq \ldots \leq p(d_{m_k i}^2)$. In practical applications, the order can be based on the actual value of the statistical distance weights.

(4) Extract the elements of the second pre-aggregation matrix that are sorted in the first preset number of rows to obtain an aggregation matrix. The number of rows in the aggregation matrix is greater than or equal to the number of predicted points of the target vehicle, and less than or equal to the number of observation points.

Use a to represent the number of selected observation points, which should satisfy $I \leq a \leq m_k$, that is, the number of selected observation points is greater than or equal to the number of predicted points of the target vehicle; extract the first row a of the left matrix of the second pre-aggregation matrix $\boldsymbol{\Delta}'$, and set the first column to 1, indicating that the observations originate from background clutter, and the aggregation matrix is obtained as

$$\boldsymbol{\Omega} = \begin{bmatrix} \omega_{10} & \omega_{11} & \omega_{12} & \cdots & \omega_{1I} \\ \omega_{20} & \omega_{21} & \omega_{22} & \cdots & \omega_{2I} \\ \vdots & \vdots & \vdots & & \vdots \\ \omega_{k0} & \omega_{k1} & \omega_{k2} & \cdots & \omega_{kI} \end{bmatrix} \tag{6.81}$$

In the formula, ω_{ji} represents the relationship between the j-th observation point and the prediction point of the i-th target vehicle; if the j-th observation point originates from the prediction point of the i-th target vehicle, then $\omega_{ji} = 1$. If the j-th observation point does not originate from the predicted point of the i-th target vehicle, then $\omega_{ji} = 0$; the first column element of $\boldsymbol{\Omega}$ is 1, indicating that the observation point originates from background clutter.

(5) Constructing multiple aggregation relationship matrices according to the aggregation matrix, each observation point in each aggregation relationship matrix is only associated with the predicted point or background clutter of one target vehicle.

Let all possible aggregation-related events at time t be $\theta(t) = \{\theta_m(t)\}_{m=1}^{n_t}$, n_t represents the number of elements in $\theta(t)$, that is, the number of aggregation-related events at time t, and $\theta_m(t)$ is

$$\theta_m(t) = \bigcap_{j=1}^{m_k} \theta_{ji}^m(t) \tag{6.82}$$

Equation (6.42) represents the possibility that in the $m-th$ aggregated related event $(m = 1, 2, \ldots, m_k)$, the m_k observation points match the predicted points of each target vehicle, and $\theta_{ji}^m(t)$ represents in the m-th aggregated related event, the observation point j originates from the event of the predicted point i of the target vehicle, and $\theta_{j0}^m(t)$ represents the event that the *j*-th observation point originates from the background clutter or false alarm.

The aggregation related events of the *j*-th observation point and the *i*-th target vehicle at time *t* are as follows:

$$\theta_{ji}(t) = \bigcup_{m=1}^{n_t} \theta_{ji}^m(t) \tag{6.83}$$

$\theta_m(t)$ can be represented by an aggregate relation matrix

$$\hat{\boldsymbol{\Omega}}[\theta_m(t)] = [\hat{\omega}_{ji}^m] = \begin{bmatrix} \hat{\omega}_{10}^m & \hat{\omega}_{11}^m & \hat{\omega}_{12}^m & \cdots & \hat{\omega}_{1I}^m \\ \hat{\omega}_{20}^m & \hat{\omega}_{21}^m & \hat{\omega}_{22}^m & \cdots & \hat{\omega}_{2I}^m \\ \vdots & \vdots & \vdots & & \vdots \\ \hat{\omega}_{m_k0}^m & \hat{\omega}_{m_k1}^m & \hat{\omega}_{m_k2}^m & \cdots & \hat{\omega}_{m_kI}^m \end{bmatrix} \tag{6.84}$$

$\hat{\boldsymbol{\Omega}}$ must ensure that there is only one non-zero value in each row of elements, it means that each observation point has only one source, which originates from the predicted point or background clutter of a target vehicle, and $i = 0$ means the 0-th target vehicle, that is, the background clutter. Except for the first column, at most one element in each column of $\hat{\boldsymbol{\Omega}}$ has a value of 1, and it indicates that the predicted point of a target vehicle has at most one observation point as its source; the first column of $\hat{\boldsymbol{\Omega}}$ can have values of multiple elements is 1, it indicates that there can be multiple observation points originating from background clutter; $\hat{\omega}_{ji}^m$ is a binary element, when the *j*-th observation point originates from the predicted point of the *i*-th target vehicle, the value is 1, otherwise the value is 0.

(6) Calculate the probability of each aggregation relationship matrix at the current moment, and filter out all the observation points associated with the prediction point of the target vehicle in the aggregation relationship matrix with the maximum probability, and perform Kalman filtering on the selected observation points, and predict the position of the predicted point of the target vehicle at the next moment.

Each aggregation relationship matrix represents an aggregation related event, that is, the relationship between all observation points in the aggregation relationship matrix at the current moment and the predicted point or background clutter of the target vehicle. By calculating the probability that each aggregation relationship matrix at the current moment appears in all the aggregation relationship matrices at the current moment, the aggregation relationship matrix with the maximum probability is obtained. The aggregation relationship matrix with the maximum probability represents the maximum probability that all the observation points in the aggregation relationship matrix at the current moment originate from the predicted point or the event set of background clutter, so the aggregation relationship matrix with the maximum probability can be used as the basis to determine the source of all observation points in the aggregation relationship matrix with the maximum probability (from the target vehicle or background clutter), and filter out all the observation points associated with the predicted points of the target vehicle in the aggregation relationship matrix with the maximum probability. Then Kalman filter is performed on the selected observation points to obtain the position of the predicted point of the target vehicle at the next moment.

Let the event at the *j*-th observation point originate from the predicted point of the *i*-th target vehicle be $\theta_{ji}(t)$, and let the event at the *j*-th observation point originate from the background clutter be $\theta_{0i}(t)$. At this time, the probability $\beta_{ji}(t)$ that the observation point is related to the predicted point aggregation of the target vehicle is:

$$\beta_{ji}(t) = P\{\theta_{ji}(t)|\mathbf{Z}^t\} \tag{6.85}$$

where, $\mathbf{Z}^t$ is all echoes that fall into the associated gate from the start of the track to the time t. From formula (6.45), it can be known that:

$$\sum_{j=1}^{m_k} \beta_{ji}(t) = 1 \tag{6.86}$$

In formula (6.45) there are:

$$P\{\theta_{ji}(t)|\mathbf{Z}^t\} = \frac{\lambda^\phi}{c^i} \prod_{j=1}^{m_k} \{N_{i_j}[Z_j(t)]\}^{\tau_j[\theta_{ji}(t)]} \prod_{i=1}^{N_\tau} (P_D)^{\delta_i}(1 - P_D)^{1-\delta_i} \tag{6.87}$$

where, P_D is the detection probability, N_τ is the number of trajectories; $N_{i_j}[Z_j(t)]$ indicates that the observation points associated with the predicted point of a target vehicle obey a Gaussian distribution; λ is the density of wrong observation points; ϕ is the number of clutter; c^i is a normalization constant; δ_i is a binary quantity, that is, whether there is an observation point interconnected with the predicted point of the target vehicle in the aggregated correlation event $\theta_{ji}(t)$. If there is an interconnected association point, δ_i is 1; if there is no interconnected association point, δ_i is 0; τ_j is a binary quantity, indicating whether the *j*-th observation point is related to a real target vehicle in the aggregated related events. Prediction point associations. τ_j is 1 if the aggregated relevant events correlate with the real target vehicle, and τ_j is 0 if the aggregated relevant events do not correlate with the real target vehicle.

Let the state estimate $\hat{X}^i_j(t|t)$ of the *j*-th observation point obtained by filtering at time p from the prediction point of the *i*-th target vehicle (the position of the prediction point of the target vehicle at the current moment) is:

$$\hat{X}^i_j(t|t) = E[\mathbf{X}^i(t)|\theta_{ji}(t), \mathbf{Z}^t] \tag{6.88}$$

The state estimation $\hat{X}^i(t|t)$ of the i-th target vehicle at time t (the position of the target vehicle at the current time) is:

$$\begin{aligned}\hat{X}^i(t|t) &= E\left[X^i(t)|Z^t\right] \\ &= \sum_{j=1}^{m_k} E\left[X^i(t)|\theta_{ji}(t), Z^t\right]P\left\{\theta_{ji}(t)|Z^t\right\} \\ &= \sum_{j=1}^{m_k} \hat{X}^i_j(t|t)\beta_{ji}(t)\end{aligned} \tag{6.89}$$

The prediction estimate $\hat{X}^i_j(t+1|t)$ of the j-th observation point at time $t+1$ originating from the prediction point of the i-th target vehicle is

$$\hat{X}^i_j(t+1|t) = A\hat{X}^i_j(t|t) + BU(t) \tag{6.90}$$

where, A and B represent the state transition matrix and the state control matrix, respectively.

The state estimate $\hat{X}^i_j(t+1|t+1)$ of the predicted point of the j-th observation point at time $t+1$ originating from the i-th target vehicle is:

$$\hat{X}^i_j(t+1|t+1) = \hat{X}^i_j(t+1|t) + K^i(t)\left[Z^t - \hat{X}^i_j(t+1|t)\right] \tag{6.91}$$

The update process of the covariance at time t is:

$$\begin{aligned}P^i(t|t) &= P^i(t|t-1) - (1-\beta_{0i})K^i(t)S^i(t)\left[K^i(t)\right]^T \\ &+ \sum_{j=1}^{m_k}\beta_{ji}(t)\left\{\begin{matrix}\hat{X}^i_j(t|t)\left[\hat{X}^i_j(t|t)\right]^T \\ -\hat{X}^i(t|t)\left[\hat{X}^i_0(t|t)\right]^T\end{matrix}\right\}\end{aligned} \tag{6.92}$$

where $K^i(t)$ is the gain matrix and $S^i(t)$ is the innovation covariance matrix.

The association method of the k-nearest neighbor joint probability data association algorithm is shown in Fig. 6.26.

(7) Repeat the first 7 steps to complete the multi-target tracking through data association.

Figure 6.27 shows the flow of the k-nearest neighbor joint probability data association algorithm.

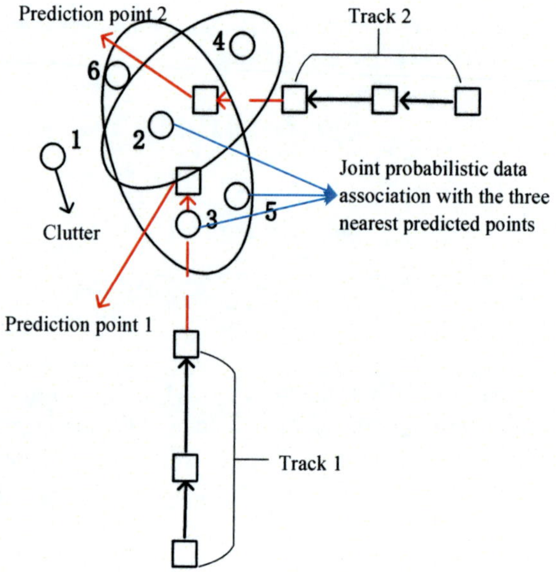

Fig. 6.26 The association method of *k*NN-JPDA algorithm

6.3.4 Experimental Design

On a four-lane highway, the k-nearest neighbor joint probability data association algorithm is tested. The test data used in the experiment is detected by the CSR-TS radar of Beijing Chuansu Microwave Technology Co., Ltd. In the experiments, the parameter k is 3.

The motion model designed here is a multi-target motion model, and there are clutter in the environment. Assuming that there are two moving targets moving in a uniform straight line with intersecting trajectories on a two-dimensional plane, the detection probability $p_d = 1$, the gate probability $p_g = 0.99$, the correlation gate threshold $\gamma = 16$, the sampling period $T = 1$s, the sampling times $n = 50$, the noise in the observation process is Gaussian White noise with known covariance and zero mean.

The initial position of target 1 is (50 m, 150 m), the initial speed is (40 m/s, 30 m/s), the initial position of target 2 is (1000 m, 150 m), and the initial speed is (−20 m/s, 40 m/s). The two moving objects are moving in a straight line at a uniform speed in the Cartesian coordinate system. 50 times of Monte Carlo simulation experiments were used to compare the results of NNDA algorithm, JPDA algorithm and kNN-JPDA algorithm.

1. Experiment 1

The multi-objective tracking results are shown in Fig. 6.28. The RMSE is compared to the algorithm accuracy, and the RMSE contrast of the three algorithms is shown in Fig. 6.29. In order to reflect the difference in the data, the average of RMSE is

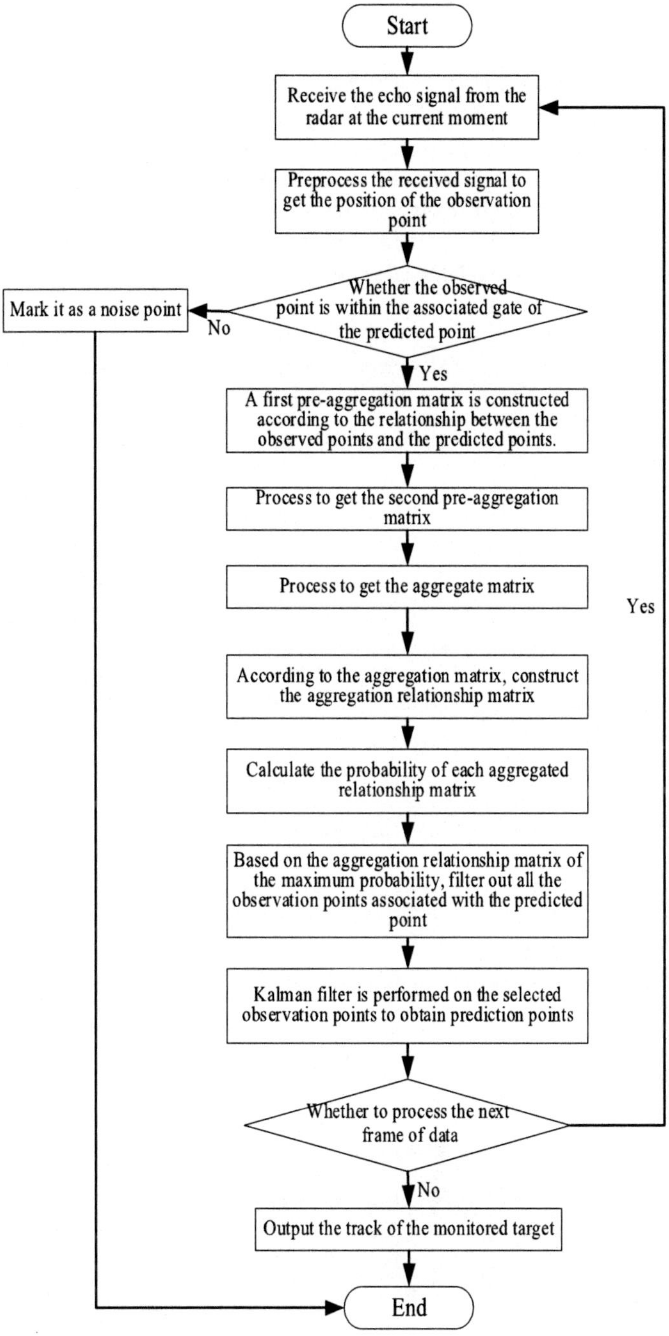

Fig. 6.27 The flowchart of *k*NN-JPDA algorithm

calculated every 10 s, and the RMSE statistics table of three algorithms are shown in Table 6.9.

As can be seen from Fig. 6.28, when the data is associated with an NNDA algorithm, a significant deviation occurs. When the JPDA algorithm is associated with the KNN-JPDA algorithm, the error is not obvious, and it cannot be clearly seen. Therefore, the introduction of RMSE to evaluate the accuracy of the algorithm, the smaller the RMSE, the higher the accuracy of the algorithm. The conclusion can be concluded from experiment 1: JPDA algorithm and the *k*NN-JPDA algorithm have no difference, all over the NNDA algorithm.

2. Experiment 2

In order to more fully understand the advantages of the *k*NN-JPDA, the comparative experiment of design algorithm runtime. Only the number of miscellaneous

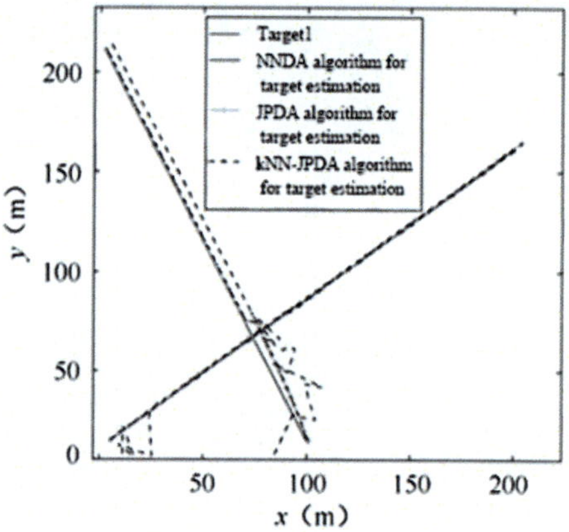

Fig. 6.28 The tracking diagram of the three algorithms

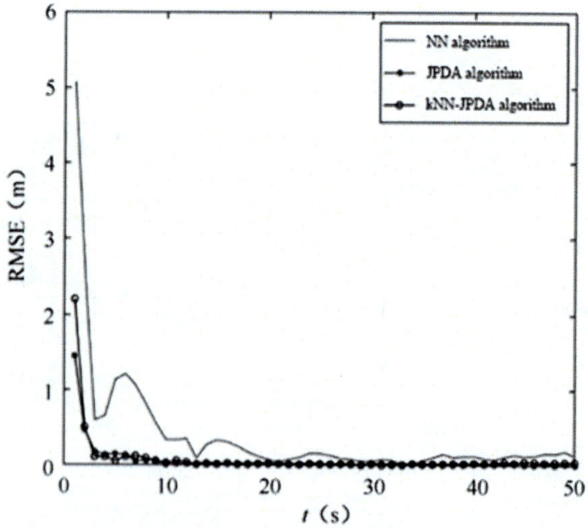

Fig. 6.29 The comparison results of RMSE

Table 6.9 The statistics of RMSE

Time (s)	RMSE (m)		
	NN	JPDA	*k*NN–JPDA
1–10	1.403	0.271	0.34
11–20	0.232	0.024	0.023
21–30	0.096	0.021	0.014
31–40	0.085	0.018	0.015
41–50	0.131	0.011	0.037

waves generated in the unit area is changed in the experiment. The three algorithms consumption, such as shown in Fig. 6.30. As can be seen from Fig. 6.30, the JPDA and the time consumption gap between the *k*NN-JPDA are large. During the definition and calculation of the algorithm, the difference between the JPDA and the *k*NN-JPDA is primarily on the number of co-inclusive events. Therefore, on the basis of the above experiment, an experiment is added, only the number of clutches generated within the unit area, the other conditions are unchanged, and the number of joint events of the two algorithms, such as shown in Fig. 6.31.

As can be seen from Fig. 6.31, the JPDA calculation rate increases with the increase in the number of clutches generated within the unit area. During the operation, the JPDA increases sharply with the increase in the number of clutters in the unit area, which leads to an extended runtime, far exceeding the *k*NN-JPDA and NNDA. Through the above experiment, the accuracy and real time of the proposed *k*NN-JPDA are better, and the superiority of the algorithm is verified below. As can be seen from the definition of this algorithm, the amount of the algorithm can be changed from the algorithm is k, and the k value is tested below.

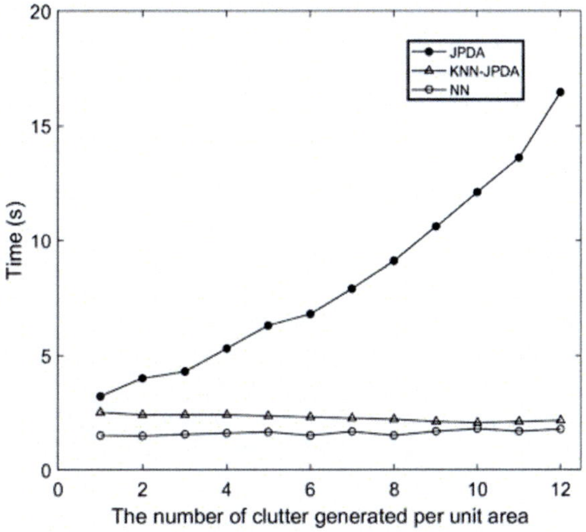

Fig. 6.30 The time consumption of the three algorithms

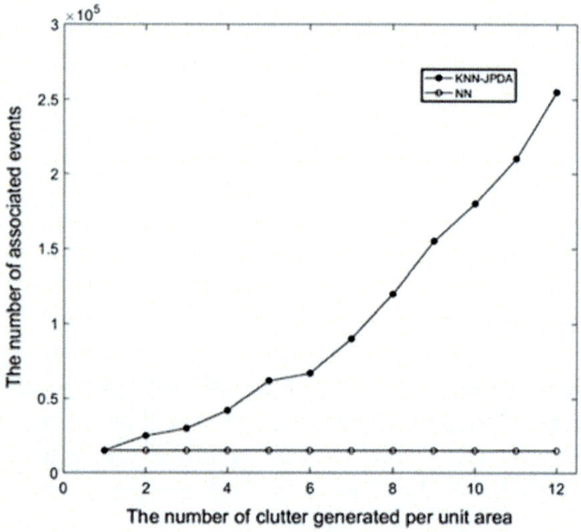

Fig. 6.31 The number of the joint events

Only k, other conditions unchanged. The time consumption when the algorithm is different k is shown in Fig. 6.32. Each 10 s calculates the average of RMSE, and the RMSE statistics table of the kNN-JPDA are shown in Table 6.10.

As k increases, the accuracy of kNN-JPDA is increased, and the time consumption increases. In practical applications, real-time is especially important, and the operation must be completed within the specified time. In the subsequent actual road test scene, k is selected to 3 or 4, which can be adjusted according to different scenes.

Actual road test uses CSR-TS radar, the test location is a four-lane road, and the test interface is shown in Fig. 6.33.

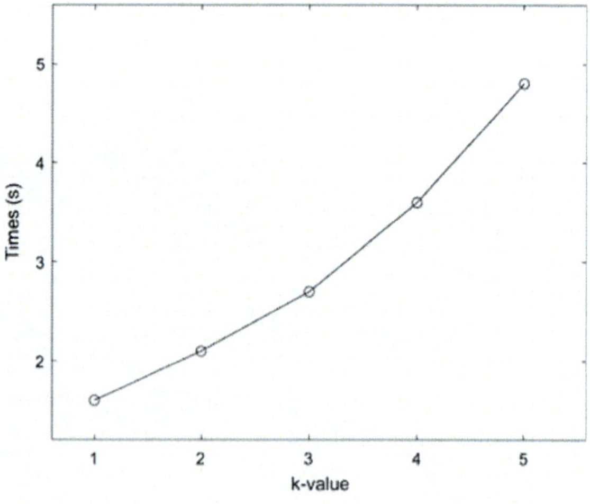

Fig. 6.32 The time consumption of kNN-JPDA with different k

Table 6.10 The comparison of RMSE

Time (s)	RMSE (m)				
	$k = 1$	$k = 2$	$k = 3$	$k = 4$	$k = 5$
1–10	0.496	0.392	0.341	0.263	0.198
11–20	0.036	0.029	0.023	0.018	0.014
21–30	0.021	0.018	0.014	0.011	0.008
31–40	0.024	0.019	0.015	0.011	0.008
41–50	0.048	0.043	0.037	0.027	0.022

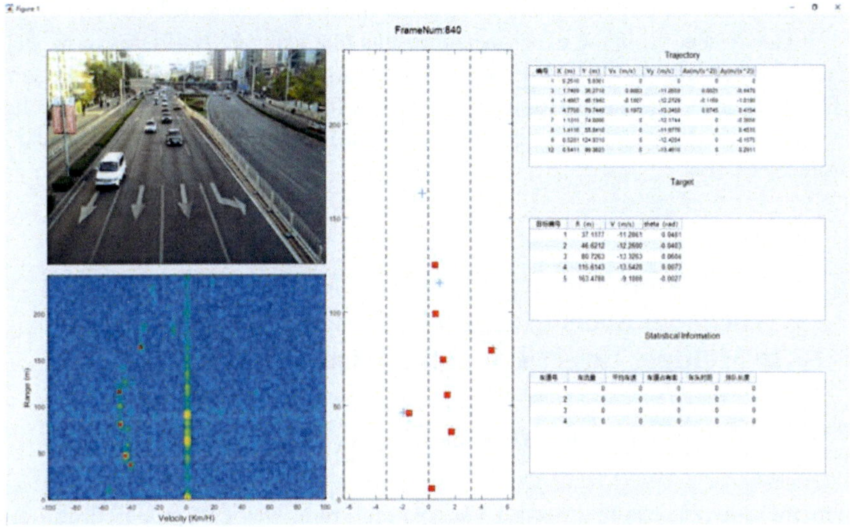

Fig. 6.33 The test interface

Data were collected under four different experimental scenarios, and the four scenarios were analyzed. The average correlation accuracy and average processing time of NNDA, JPDA and *k*NN-JPDA are counted, and the performance comparison of the three algorithms is shown in Table 6.11.

Table 6.11 The performance comparison of the three algorithms

Algorithm	Average processing time (ms)	Average correlation accuracy (%)
NNDA	27	70.86
JPDA	61	96.64
*k*NN–JPDA	34	95.87

The correct correlation is:

$$\eta = \frac{R}{R + W} \tag{6.93}$$

In the formula, η is the correlative correct rate, R is the number of sample points associated with the correct sample point, W is the number of sample points associated. As can be seen from Table 6.11, the average correlation rate of the kNN-JPDA can reach over 95%, and the average processing time is about 34 ms. Compared to kNN-JPDA, although the NNDA runs shorter, its correlation is low and cannot be applied in actual scenarios. The JPDA has a high correlation rate, but its processing time is long, and it is low efficiency, and it is impossible to apply in actual scenes.

It can be seen from the experimental results that using kNN-JPDA can be well-larger, and accurately reflect the location of the target, motion state, etc. At the same time, it has the real-time performance to achieve the expected state, which is in line with the multi-target tracking requirements of millimeter wave surveillance radar in traffic scenarios.

6.4 Convex Variational Inference for Multi-hypothesis Fractional Belief Propagation Based Data Association in Multiple Target Tracking Systems

6.4.1 Multiple-Hypothesis Tracking

The multi-hypothesis multi-target tracking algorithm was first proposed by Reid, which is based on the "full neighbor" optimal filter and the aggregation concept proposed by Bar-Shalom. And developed the concept of association hypothesis through three consecutive refinements (numbering, configuration and assignment) of the result space. The algorithm mainly includes: the formation of poly, the generation of "hypothesis", and the probability calculation of each hypothesis. The "hypothesis" proposed by the multi-hypothesis algorithm is almost the same as the concept of joint events proposed by Bar-Shalom. The main differences are: (1) For each echo, not only the possibility of false alarms but also the possibility of the emergence of new targets are considered. (2) We treat the hypothesis at time k as the result of the interconnection between a certain hypothesis at time $k-1$ and the current data set.

We assume that $\boldsymbol{q}_l^k$ is the set of interconnected hypotheses from the start time to the time k, and the measurement set at each time is defined as $Z(k) = \{Z_i(k)\}_{i=1}^{m_k}$. In multi-target tracking, it is usually assumed that in each sensor scan, each target produces at most one measurement, but it is not known which target the measurement comes from and which measurement is a false alarm. In principle, as long as the posterior probability distribution $p(\boldsymbol{q}_l^k|Z^k)$ is determined, the multi-target tracking problem can be solved. In statistical estimation theory, it has been well proved that the minimum probability of error estimation can be given by the maximum a posteriori (MAP) estimator. In addition, the use of particle filtering was proposed in [25] to reduce the complexity of calculating the posterior probability. The key to MHT is how to perform data association calculations before receiving more data.

6.4.1.1 Hypothesis-Oriented MHT

From the perspective of calculation and real-time performance, we need to establish a recursive formula for $P(\boldsymbol{q}^k|Z(k))$. First, we define a global hypothesis $\boldsymbol{q}_l^k$ at time k-th. The definition $\boldsymbol{q}_l^k$ is derived from the hypothesis $\boldsymbol{q}_{m(l)}^{k-1}$ at the previous time and the event q_l^k at the current time, that is $\boldsymbol{q}_l^k = \boldsymbol{q}_{m(l)}^{k-1}, q_l^k$. We denote q_m^k as a specific set of hypotheses mapping $\boldsymbol{q}_{m(l)}^{k-1}, Z(k)$ to q_l^k, that is q_m^k represents the set of all hypotheses $\boldsymbol{q}_{m(l)}^{k-1}$ at time k. In addition, for the convenience of reading and presentation, we introduce the variable ζ and ζ = |Z(k)|.

$$\begin{aligned}P(q_j^k|Z(k)) &= P(q_j^k|Z(k),\zeta)\\ &= \frac{P(q_j^k|Z(k-1),\zeta)P(Z(k)|Z(k-1),q_j^k,\zeta)}{P(Z(k)|Z(k-1),\zeta)}\end{aligned} \tag{6.94}$$

In addition, the parameters ζ can be known through the established q_j^k, so (1) can be simplified to:

$$P(Z(k)|Z(k-1),q_j^k,\zeta) = P(Z(k)|Z(k-1),q_j^k) \tag{6.95}$$

The second molecular factor in formula (6.1) can be expanded as follows:

$$\begin{aligned}P(q_j^k|Z(k-1),\zeta) &= P(q_j^k|Z(k-1),\zeta,q_j^{k-1})P(q_j^k|Z(k-1),\zeta)\\ &= \frac{\begin{array}{c}P(q_j^k|Z(k-1),\zeta,q_j^{k-1})P(\zeta|q_j^{k-1},Z(k-1))\\ P(q_j^{k-1}|Z(k-1))\end{array}}{P(\zeta|Z(k-1))}\\ &= \frac{P(q_j^k|Z(k-1),q_j^{k-1})P(q_j^{k-1}|Z(k-1))}{P(\zeta|Z(k-1))}\\ &= \frac{P(q_j^k|q_j^{k-1})P(q_j^{k-1}|Z(k-1))}{P(\zeta|Z(k-1))}\end{aligned} \tag{6.96}$$

The denominator of formula (1) can be written as:

$$P(Z(k)|Z(k-1),\zeta) = \frac{P(Z(k)|Z(k-1))}{P(\zeta|Z(k-1))} \tag{6.97}$$

substitute (6.95–6.97) into (6.94) to get:

$$P(q_j^k|Z(k)) = \frac{P(Z(k)|Z(k-1), q_j^k)P(q_j^k|q_j^{k-1})P(q_j^{k-1}|Z(k-1))}{P(Z(k)|Z(k-1))} \tag{6.98}$$

This is to recursively express the global hypothesis $P(q_j^k|Z(k))$ as $P(q_j^{k-1}|Z(k-1))$ and scan data $Z(k)$ at the current moment.

In fact, the hypothetical probability at time k can be calculated iteratively at time $k-1$. In addition, there is no need to consider inconsistent global hypotheses $Z(k)$ because their posterior probability is 0.

6.4.1.2 Track-Oriented MHT

Although the above (6.97) can solve the problem, the recursive method in (6.97) is tricky in terms of global assumptions in a large spatial sense. Fortunately, under some simplified assumptions, that is, assuming that the number of target occurrences and the number of false alarms per scan obey the Poisson distribution, the posterior probability of the overall hypothesis $P(\boldsymbol{q}^k|Z(k))$ can be expressed as a local (or trajectory) hypothesis related to the product of $\boldsymbol{q}^k$. Moreover, the Poisson distribution assumption is reasonable in many cases.

Now we show the details in (6.97). First, for $t = \{1, 2, ..., m(l)\}$ define the following marker variables related to the event q_l^k:

$$\tau_t = \tau_t(q_l^k) = \begin{cases} 1, \text{Z}_\text{i}\text{(k)from the confirmed measurements} \\ 0, \text{ otherwise} \end{cases} \tag{6.99}$$

$$\upsilon_t = \upsilon_t(q_l^k) = \begin{cases} 1, \text{ Z}_\text{i}\text{(k) from the new targets} \\ 0, \text{ otherwise} \end{cases} \tag{6.100}$$

$$\chi_t = \chi_t(q_l^k) = \begin{cases} 1, \text{ detected track } t \text{ at } k\text{ - th} \\ 0, \text{ otherwise} \end{cases} \tag{6.101}$$

The probability of each hypothesis $P(\boldsymbol{q}_l^k|Z(k))$ can be calculated by Bayesian method:

$$\begin{aligned} P(\boldsymbol{q}_l^k|Z(k)) &= P\{q_l^k, \boldsymbol{q}_{m(l)}^k|Z(k), Z(k-1)\} \\ &= \frac{1}{c}P\{Z(k)|q_l^k, \boldsymbol{q}_{m(l)}^{k-1}, Z(k-1)\} \\ &\quad \cdot P\{q_l^k|\boldsymbol{q}_{m(l)}^{k-1}, Z(k-1)\}P\{\boldsymbol{q}_{m(l)}^{k-1}|Z(k-1)\} \end{aligned} \tag{6.102}$$

where c is the normalization constant factor. $P\{\boldsymbol{q}_{m(l)}^{k-1}|Z(k-1)\}$ represents the probability of the global hypothesis at time $k-1$. Then:

$$P\{q_l^k|\mathbf{q}_{m(l)}^{k-1}, Z(k-1)\} = \frac{\phi!\upsilon!}{m_k!}\mu_F(\phi)\mu_N \prod_t \frac{(P_D^t)^{\delta_t}(1-P_D^t)^{\delta}}{(P_\chi^t)^{\chi_t}(1-P_\chi^t)^{1-\chi_t}} \tag{6.103}$$

Among them, $\mu_F(\phi)$ and $\mu_N(\upsilon)$ respectively represent the prior probability function of false measurement and new geometric features, and P_D^t and P_χ^t respectively represent the detection and termination probability of the track t.

The probability of occurrence of measurement and false alarm obeys the Poisson distribution, and the parameters are λ_F and $\lambda_{N'}$ respectively, then:

$$\begin{aligned} P(q_m^k|Z(k)) = \frac{1}{c}\lambda_{N'}^{\upsilon}\lambda_F^{\phi}\prod_{i=1}^{m_k}[N_{t_i}[z_i(k)]]^{\tau_i}\{\prod_t (P_D^t)^{\delta_t} \\ (1-P_D^t)^{1-\delta_t}(P_\chi^t)^{\chi_t}(1-P_\chi^t)^{1-\chi_t}P\{q_{m(l)}^{k-1}|Z(k-1)\} \end{aligned} \tag{6.104}$$

Equation (6.104) is of fundamental importance because it incorporates the global hypothesis score into the (dimensionless) tracking score. The path-oriented MHT (TO-MHT) recursion given by it can efficiently implement the hypothesis-oriented MHT (HO-MHT) recursion given in (6.97). In addition, under the assumption that both the target and the clutter obey the Poisson distribution, these two recursions are equivalent. Therefore, both methods include MAP estimation for global hypotheses, followed by maximum a posteriori (MAP) or minimum mean square error (MMSE) processing based on data association solutions. In the case of linear Gaussian, MAP and MMSE filtering are equivalent, which is given by the well-known Kalman filter. It needs to be pointed out that TO-MHT recursion depends on the target is uneven (Poisson) prior-distributed.

6.4.2 *Multi-hypothesis Fractional Belief Propagation*

6.4.2.1 Probabilistic Graphical Model and Variational Inference

The Graphical Model is a bridge between probability theory and graph theory, that is, a complex system is composed of simple parts. A complex radar multi-target tracking system can be decomposed into several scans. At each moment, the source of measurement is mainly divided into: the measurement comes from the target, and the measurement is clutter.

In the actual detection scene, when the predicted positions of multiple targets are close, that is, the relevant gates overlap. As shown in Fig. 6.34, at this time, it is necessary to comprehensively consider the target sources of all measurements (Z_{t1}, Z_{t2}, Z_{t3}). When there are only two targets, and only one measurement in the intersecting gate, with the moderate number of hypotheses generated by the MHT algorithm, the relatively ideal correlation effect can be achieved, ensuring the computational complexity and the correlation accuracy simultaneously; However, in actual

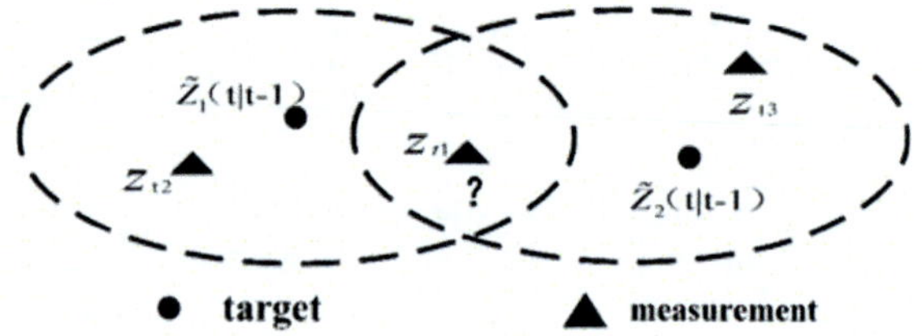

Fig. 6.34 Example of interconnection events at time *t*

multi-close target detection scenarios, as the number of targets increases and the predicted position approaches, the number of intersecting gates will also increase. Accordingly, the number of measurements in the intersecting gate increases, and the ambiguity of the source of each measurement tends to increase. The corresponding number of hypotheses (and the tracks in these hypotheses) that the MHT system can generate advance obviously at an exponential rate. The MHT algorithm can no longer keep the balance between computational complexity and correlation accuracy, that is, it doesn't meet the requirements for the accuracy of the algorithm in the radar tracking system.

As shown in Fig. 6.35, the classic MHT track branching structure diagram can provide a convenient mechanism for implementing delayed decision logic and show how to define a convenient structure for track construction. Using this structure, a family is defined as a set of trajectories with a common root node, or we define a family (all trajectories from a single ancestor or root node) which can also be regarded as a target tree. Each branch represents a different data association hypothesis for a single target, and a node is defined as a point where a track forms two or more branches. Because each branch track in the family (target tree) has at least one common node (root node), these tracks are incompatible with each other and can represent at most one target. The black dashed line shows an example of a global hypothesis, that is, a single target hypothesis is selected for each track (branch) from the tree, where each measurement in each scan is used exactly once.

Typical probabilistic graphical models mainly include Bayesian networks and Markov random fields. Bayesian network is a directed graph model, which is used to express the causal relationship between random variables; while Markov random field is an undirected graph model, which is used to express the probability distribution reasoning of random variables or the relationship between random variables soft constraint relations; a new structure can be constructed from the first two networks called factor graphs.

In the undirected tree graph, the belief propagation algorithm (BP) can be used for precise inference. The foundation of the BP algorithm is based on MRF (Markov Random Field). MRF is a conditional probability model, which can be considered as a generalization of Markov chain, which is a very effective description for the correlation of all nodes in the field. For example, we suppose that $\psi_i(x_i) = p(x_i)$, $\psi_{i,j} = p(x_j|x_i)$, and $\psi_{i,j}$ are also called discontinuous costs between adjacent nodes, reflecting the compatibility between node variables x and y. The iterative update equation is as follows:

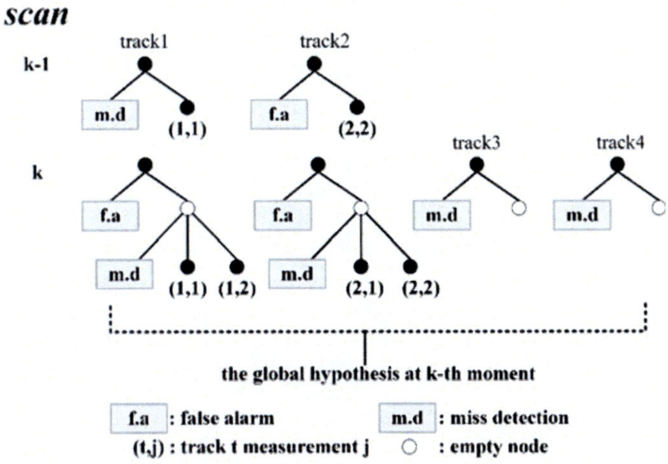

Fig. 6.35 Global hypothesis formed at scan k

$$\mu_{i\to j}(x_j) \propto \sum_{\theta_i} \psi_{i,j}(x_i, x_j)\psi_i(x_i) \prod_{(j',i)\in\xi, j'\neq j} \mu_{j'\to i}(x_i) \tag{6.105}$$

It is also called the sum-product algorithm. When the synthesis is replaced with the maximum value, the max-product BP algorithm is obtained, which provides the joint state of the MAP of all nodes in the tree structure graph. When the sum-product BP algorithm converges, the marginal distribution of the vertex is:

$$p(x_j) \propto \psi_j(x_j) \prod_{(j,i)\in\xi} \mu_{i\to j}(x_j) \tag{6.106}$$

In a Markov chain, if all nodes are independent and identically distributed and obey Gaussian distribution, BP is equivalent to Kalman filter. Similarly, if all nodes are discrete, BP is equivalent to using the forward–backward algorithm to infer the HMM. The BP algorithm unifies these algorithms and extends the structure from a chain structure to a tree structure.

For the specific graph model structure, if the corresponding BP algorithm converges, the accurate marginal probabilities can be restored by optimizing the convex function called Gibbs free energy. In the case of a single vertex (or if all variables are combined into a single vertex), the specific content is expressed as follows, where q represents the estimated value of the marginal probability distribution, also known as the belief:

The Gibbs free energy variational problem of a random variable x can be written as:

$$\min_{q(x)} imise -H(x) - \mathrm{E}[\log \psi_x] \tag{6.107}$$

$$subject\ to\ q(x) \geq 0, \sum_x q(x) = 1 \tag{6.108}$$

where:

$$\begin{cases} \mathrm{E}[\log \psi(x)] \triangleq \sum_x q(x) \log \psi(x) \\ H(x) = -\mathrm{E}[\log q(x)] \end{cases} \tag{6.109}$$

The optimal solution is:

$$q(x) = \frac{\psi(x)}{\sum_{x'} \psi(x')} \propto \psi(x) \tag{6.110}$$

Similar expressions apply for continuous random variables, replacing sums with integrals.

Equation (6.107) can be recognized as a measure of the Kullback–Leibler (KL) divergence between $q(x)$ and the unnormalized distribution $\psi(x)$. If it is a tree structure, $H(x)$ can be decomposed into:

$$H(x) = \sum_{v \in V} H(x_v) - \sum_{(i,j) \in \xi} I(x_i; x_j) \tag{6.111}$$

where:

$$I(x_i; x_j) = H(x_i) + H(x_j) - H(x_i, x_j) \tag{6.112}$$

$I(x_i; x_j)$ is the mutual information between x_i and x_j. Accordingly, the variational problem can be written as:

$$\begin{aligned} \underset{q(x_v), q(x_i, x_j)}{\text{minimise}} \ & - \sum_{v \in V} \{H(x_v) + \mathrm{E}[\log \psi_v(x_v)]]\} \\ & - \sum_{(i,j) \in \xi} \{-I(x_i; x_j) + \mathrm{E}[\log \psi_{i,j}(x_i, x_j)]\} \end{aligned} \tag{6.113}$$

$$subject\ to\ \ q(x_i, x_j) \geq 0\ \forall (i, j) \in \xi, \forall x_i, x_j \tag{6.114}$$

$$\sum_{x_v} q(x_v) = 1\ \ \forall v \in V \tag{6.115}$$

$$\sum_{x_j} q(x_i, x_j) = q(x_i)\ \ \forall (i, j) \in \xi \tag{6.116}$$

$$\sum_{x_i} q(x_i, x_j) = q(x_j) \ \forall (i, j) \in \xi \tag{6.117}$$

Equation (6.113) Also known as Bayesian free energy. In the cycle graph, it is not same as Gibbs free energy. It is still a general approximate solution for tree graph. The feasible set described by formulas (6.113)-(6.117) is accurate, that is, any effective solution can be obtained through an effective joint distribution, and any effective joint distribution can be mapped to a feasible solution.

6.4.2.2 Multi-hypothesis Fractional Belief Propagation Data Association

The probabilistic graphical model mainly expresses and calculates the joint probability distribution between multiple variables by decomposing the implicit connections between variables. From Fig. 6.35, we can find that there is such a probability connection between the track formed at scan k and the measurement at scan $k-1$. The main purpose of MHT is to list all possible hypotheses and find the largest of them.

We build a probabilistic graph model based on MHT. As a multi-hypothesis multi-target tracking algorithm, in simple terms, MHT is to track all assumptions for all appearing targets, and "prune" those false assumptions through multi-frame tracking. During the processing, not only the hypothesis tracking at a single scan must be considered, but also the hypothesis set before that time must be combined. Therefore, now we define $\boldsymbol{\alpha}(k-1) = \{\alpha_1(k-1), \alpha_2(k-1), ..., \alpha_t(k-1)\}$ to represent the formed track at time $k-1$. At time k, the trajectory is $\boldsymbol{\alpha}(k) = \{\alpha_1(k), \alpha_2(k), ..., \alpha_t(k), \alpha_{t+1}(k), ..., \alpha_{t+m_t}(k)\}$, where $\alpha_{t+m_t}(k)$ represents the new trajectory hypothesis formed based on the m_t measurements received at time k, as shown in the red dotted circle in Fig. 6.36. The dotted line in the figure indicates that the hypothetical track is not established, and the solid line represents the formed track.

At each k-th scan, we need to find the correlation state between the formed track and the detected measurement. In order to combine MHT with the probabilistic graph model, two association indicators θ_i and δ_j are introduced. For each track $i \in \{1, 2, ..., n\}$, $\theta_i \in \{0, 1, 2, ..., m_t\}$ is a track association indicator indicated to the track i that is assumed to be related to the measurement $z(j)$. Similarly, for each measurement $j \in \{0, 1, 2, ..., m_t\}$, $\delta_j \in \{0, 1, 2, \ldots, n\}$ is a measurement association indicator indicated to the measurement, which is assumed that it is correlated with trajectory i (0 means that the detected measurement is a false alarm or a new trajectory).

First we define $G = (V, E)$ as an undirected graph, as shown in Fig. 6.37, including vertex $v \in V$ and edge $e \in \xi \in V \times V$. The leaf nodes of the first, second, and third layers in the probability graph respectively represent the track α_i (all hypotheses) formed in the detection area at k, the track correlation indicator θ_i, the measurement correlation indicator δ_j, and the edge $\mu_{i\to j}$ indicates the association

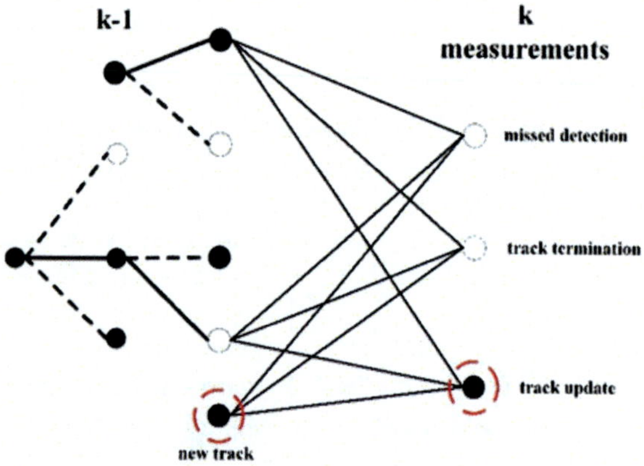

Fig. 6.36 The diagram of MHT association

status between the track association indicator θ_i and the measurement association indicator δ_j.

The probabilistic graph model clearly shows the redundancy in the global hypothesis set. However, this redundancy can implicitly ensure that each measurement corresponds to at most one trajectory, and each trajectory corresponds to at most one measurement.

The main purpose in this paper is to study the data association between trajectory and measurement at a single scan, which involves multiple targets. First, we assume that there are n targets in the detection range, and each target is associated with at most one measurement, and each measurement originates from at most one target. Among them, the probability of false alarms obeys Poisson distribution. And we define the joint state of n trajectories as $\boldsymbol{\alpha} = \{\alpha_1, \alpha_2, ..., \alpha_n\}$, and the joint state of the measurement at t as $\mathbf{Z} = \{ Z_1,Z_2,...,Z_{m_t} \}$. Assuming that the prior information of each target is independent of each other, the joint distribution of $\boldsymbol{\alpha}$ is:

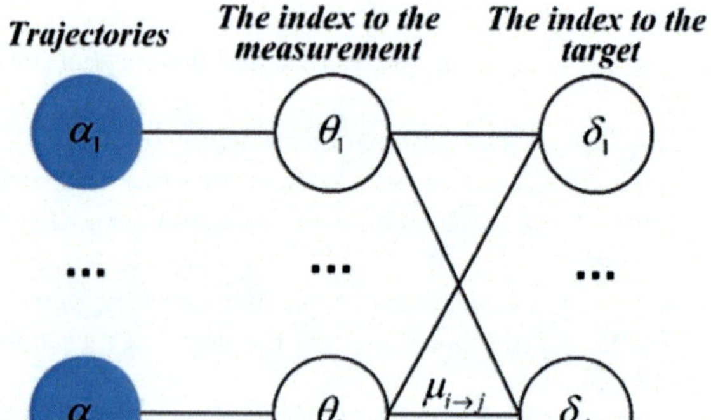

Fig. 6.37 Probabilistic graphical model

$$p(\boldsymbol{\alpha}) \propto \prod_{i=1}^{n} \psi_i(\alpha_i) \tag{6.118}$$

where $\psi_i(\alpha_i)$ represents the joint prior probability of each α_i.

Next, we define the probability that each trajectory i is detected as $P^d(\alpha_i)$, the correlation model between the target and the measurement is $p(Z|\alpha_i)$, and the parameter of the Poisson distribution is $\lambda(Z)$. Combined with the correlation indicator variables θ_i and δ_j, then the joint distribution between the measurement and indicator variables is:

$$\begin{aligned} p(Z, \boldsymbol{\theta}, \boldsymbol{\delta}|\boldsymbol{\alpha}) \propto & \left\{ \prod_{i|\theta_i>0} P^d(\alpha_i) p\big(Z_{\theta_i}|\alpha_i\big) \right\} \\ & \times \left\{ \prod_{i|\theta_i=0} [1 - P^d(\alpha_i)]\} \times \{ \prod_{j|\delta_j=0} \lambda(Z_j)\} \times \psi(\boldsymbol{\theta}, \boldsymbol{\delta}) \right\} \end{aligned} \tag{6.119}$$

The posterior probability distribution is:

$$p(\boldsymbol{\alpha}, \boldsymbol{\theta}, \boldsymbol{\delta}|Z) \propto p(\boldsymbol{\alpha}) \times p(Z, \boldsymbol{\theta}, \boldsymbol{\delta}|\boldsymbol{\alpha}) \tag{6.120}$$

Because the measurement is a constant in each scan, formula (6.118) is divided by $\prod_{j=1}^{m_t} \lambda(Z_j)$. So:

$$p(\boldsymbol{\alpha}, \boldsymbol{\theta}, \boldsymbol{\delta}|Z) \propto \prod_{i=1}^{n} \{\psi_i(\alpha_i) \cdot \psi_i(\alpha_i, \theta_i) \cdot \prod_{j=1}^{m_t} \psi_{i,j}(\theta_i, \delta_j)\} \tag{6.121}$$

where:

$$\psi_i(\alpha_i, \theta_i) = \begin{cases} \dfrac{P^d(\varepsilon_i) p(Z_j|\alpha_i)}{\lambda(Z_j)}, \theta_i = j > 0 \\ 1 - P^d(\alpha_i), \ \ \theta_i = 0 \end{cases} \tag{6.122}$$

$$\psi(\boldsymbol{\theta}, \boldsymbol{\delta}) = \begin{cases} 1, \ \theta_i \text{ correlate to } \delta_j \\ 0, \ \text{otherwise} \end{cases} \tag{6.123}$$

At multiple time steps, the marginal distribution $p_i(\alpha_i)$ of each target is calculated at first (the marginal distribution can be fitted by Gaussian distribution), and then proceed to the next time step, using the product of these approximate marginal distributions to approximate joint prior distribution.

On the tree structure diagram shown in Fig. 6.37, belief propagation (BP) can be used to make the best inference. BP is mainly carried out by passing messages between neighboring nodes. Next, the main purpose is to combine data association with BP algorithm. A simplified version of the BP equation has been proposed and proved. First, we define $q_i(x_i)$ to represent the belief (approximate probability) of the variable x_i, and $q_{i,j} = q(\theta_i = j) = q(\delta_j = i)$ to represent the belief of the association between the trajectory i and the measurement j. In addition, if $j = 0$, it means that the target i is missed; if $i = 0$, it means that the measurement j is a false alarm.

The variational problem of single-time radar scanning is mainly solved by minimizing the target (6.124):

$$\underset{q_{i,j}}{\text{minimise}} \sum_{i=1}^{n}\sum_{j=0}^{m_t} q_{i,j} \log \frac{q_{i,j}}{\psi_i(\theta_i = j)} + \sum_{j=1}^{m_t} q_{0,j} \log q_{0,j} - \sum_{i=1}^{n}\sum_{j=1}^{m_t} (1 - q_{i,j}) \log(1 - q_{i,j}) \tag{6.124}$$

subject to:

$$\sum_{j=0}^{m_t} q_{i,j} = 1 \ \forall i \in \{1, 2, \ldots, n\} \tag{6.125}$$

$$\sum_{i=0}^{n} q_{i,j} = 1 \ \forall j \in \{1, 2, \ldots, m_t\} \tag{6.126}$$

$$0 \le q_{i,j} \le 1 \tag{6.127}$$

Among them, the constraints of (6.125) and (6.126) are called consistency constraints, and they are necessary conditions for the solution to correspond to the distribution of effective joint correlation events. If the correct score coefficient $\gamma \in [-1, 1]$ is incorporated in the last item of the objective (6.124) (excluding false alarms and missed detections), then the optimal value of the optimal objective function is the same as the goal of the Gibbs free energy. In the formula that combines false alarms and missed detections, the fractional free energy (FFE) is:

$$\begin{aligned} F_B^{\gamma}(q_{i,j}) = &\sum_{i=1}^{n}\sum_{j=0}^{m_t} q_{i,j} \log \frac{q_{i,j}}{\psi_i(a_i = j)} \\ &+ \gamma \sum_{j=1}^{m_t} q_{0,j} \log q_{0,j} \\ &- \gamma \sum_{i=1}^{n}\sum_{j=1}^{m_t} (1 - q_{i,j}) \log(1 - q_{i,j}) \end{aligned} \tag{6.128}$$

When $\gamma = 1$, (6.128) is Bayesian free energy (BFE). Although the "correct" value of γ cannot be determined for any particular problem, it is found that the coefficient $\gamma \in [0, 1]$ can improve belief. In fact, the value can be selected in advance based on the problem. When the fractional coefficient $\gamma \in [-1, 1]$, the convexity of (6.124) can be retained on an appropriate subset.

Now we consider how to apply these results to solve the single-scan problem with dynamic trajectory state $\boldsymbol{\alpha}_i$, as shown in Fig. 6.37. The Bayesian free energy function can be constructed by combining $q_i(\theta_i)$ and $q_i(\alpha_i, \theta_i)$ to solve the objective (6.124):

$$\begin{aligned} F_B(q_i(\alpha_i), q_i(\alpha_i, \theta_i), q_{i,j}) = &-\sum_{i=1}^{n}\{H(\alpha_i) + \mathrm{E}[\log \psi_i(\alpha_i)] \\ &+ H(\theta_i|\alpha_i) + \mathrm{E}[\log \psi_i(\alpha_i, \theta_i)]\} \\ &+ \sum_{j=1}^{m_t} q_{0,j} \log q_{0,j} - \sum_{i=1}^{n}\sum_{j=1}^{m_t}(1 - q_{i,j}) \log(1 - q_{i,j}) \end{aligned} \tag{6.129}$$

subject to:

$$q_i(\alpha_i, \theta_i) \geq 0, \;\; q_i(\alpha_i) \geq 0 \tag{6.130}$$

$$\sum_{\alpha_i}\sum_{\theta_i=0}^{m_t} q_i(\alpha_i, \theta_i) = 1 \tag{6.131}$$

$$\sum_{\theta_i=0}^{m_t} q_i(\alpha_i, \theta_i) = q_i(\alpha_i), \;\; \forall i \tag{6.132}$$

$$q_{i,j} = \sum_{\alpha^i} q_i(\alpha_i, j) \; \forall i, j \tag{6.133}$$

$$q_{i,j} \geq 0 \; \forall i, j \tag{6.134}$$

$$\sum_{i=0}^{n} q_{i,j} = 1 \; \forall j \tag{6.135}$$

In order to ensure that the objective function can converge to the optimal value, now we need to make some changes on the basis of (6.129) to construct a convex free energy function:

$$F_B^{\gamma,\beta}(q_i(\alpha_i), q_i(\alpha_i, \theta_i), q_{i,j}) = -\sum_{i=1}^{n}\{H(\alpha_i) + \mathrm{E}[\log \psi_i(\alpha_i)]\}$$

$$+\beta\sum_{j=1}^{m_t} q_{0,j}\log q_{0,j}$$
$$-\gamma\sum_{i=1}^{n}\sum_{j=1}^{m_t}(1-q_{i,j})\log(1-q_{i,j})$$
$$-\sum_{i=1}^{n}\{H(\theta_i|\alpha_i)+\mathrm{E}[\log\psi_i(\alpha_i,\theta_i)]\} \tag{6.136}$$

The constraints are the same as (6.130–6.135), and the coefficients are $\gamma\in[0,1)$ and $\beta\in(0,1]$. This re-weighting method is to obtain convex free energy that satisfies the solution conditions. In order to optimize (6.136) more conveniently, now we consider decomposing it as follows:

$$\begin{aligned}F_B^{\gamma,\beta}(q_i(\alpha_i),q_i(\alpha_i,\theta_i)) &= f(q_i(\alpha_i),q_i(\alpha_i,\theta_i))\\ +h_1(q_i(\alpha_i),q_i(\alpha_i,\theta_i)) &+ h_2(q_i(\alpha_i),q_i(\alpha_i,\theta_i))\end{aligned} \tag{6.137}$$

where:

$$\begin{aligned}&f(q_i(\alpha_i),q_i(\alpha_i,\theta_i))\\ &=-\sum_{i=1}^{n}\{k_{f,\alpha}H(\alpha_i)+\mathrm{E}[\log\psi_i(\alpha_i)]\}\\ &-\sum_{i=1}^{n}\{k_f H(\alpha_i,\theta_i)+\mathrm{E}[\log\psi_i(\alpha_i,\theta_i)]\}\end{aligned} \tag{6.138}$$

$$\begin{aligned}h_1(q_i(\alpha_i),q_i(\alpha_i,\theta_i)) &= -\sum_{i=1}^{n}\{k_{1,\alpha}H(\alpha_i)+k_1(\alpha_i,\theta_i)\}\\ &+\beta\sum_{j=1}^{m_t}q_{0,j}\log q_{0,j}-\gamma\sum_{i=1}^{n}\sum_{j=1}^{m_t}(1-q_{i,j})\log(1-q_{i,j})\end{aligned} \tag{6.139}$$

$$h_2(q_i(\alpha_i),q_i(\alpha_i,\theta_i)) = -\sum_{i=1}^{n}\{k_{2,\alpha}H(\alpha_i)+k_2H(\alpha_i,\theta_i)\} \tag{6.140}$$

Among them, (6.137) and (6.139) do not depend on $q_{i,j}$, because they depend on $q_i(\alpha_i,\theta_i)$ bound by (6.133–6.135). The following is a description of the convexity of (6.138–6.140):

(a) If $k_f>0$ and $k_{f,\alpha}>0$, then f is strictly convex.
(b) If $k_1\geq 0,\beta\geq 0$ and $k_{1,\alpha}+k_1\geq\gamma$, then h_1 is strictly convex.
(c) If $k_2\geq 0$ and $k_2{+}k_{2,\alpha}\geq 0$, then h_2 is strictly convex.

Using the Primal–Dual Coordinate Ascent (PDCA) algorithm, the convex free energy in (6.137) can be solved by optimizing the following objectives:

$$\begin{aligned} \text{minimise } & \sum_{i=1}^{n}\sum_{j=0}^{m_t} q_{i,j}\log\frac{q_{i,j}}{\varpi_{i,j}} + \tilde{\beta}\sum_{j=1}^{m_t} q_{0,j}\log q_{0,j} \\ & -\tilde{\gamma}\sum_{i=1}^{n}\sum_{j=1}^{m_t}(1-q_{i,j})\log(1-q_{i,j}) \end{aligned} \tag{6.141}$$

subject to (6.134–6.135), and $\sum_{j=0}^{m_t} q_{i,j} = 1, \forall i$.

where:

$$\begin{cases} \tilde{\beta} = \dfrac{\beta}{k_f + k_1} \\ \tilde{\gamma} = \dfrac{\gamma}{k_f + k_1} \end{cases} \tag{6.142}$$

$$\varpi_{i,j} = \psi_i(\theta_i = j) = \int \psi_i(\alpha_i)\psi_i(\alpha_i, \theta_i = j)d\alpha_i \tag{6.143}$$

The flow of the multi-hypothesis fractional belief propagation algorithm for a single scan is shown in the figure below, where ε is the coefficient of $q_{i,0}$ and $\log\frac{q_{i,o}}{\varpi_{i,0}}$:

The multi-hypothesis fractional belief propagation for single scan
Input: $n, m_t \in N, \gamma \in [0, 1)$; $\varpi_{i,j} \forall i \in \{0, 1, ..., n\}, j \in \{0, 1, ..., m_t\}$ $\varepsilon \in (0.5, \infty) \cap [\gamma, \infty)$; $\beta \in (0.5, \infty) \cap [\gamma, \infty]$
Initialise: $k = -1 - \gamma + \beta + \varepsilon$ $\varsigma_{i,j} = 1 \ \forall i \in \{0, 1, ..., n\}, j \in \{0, 1, ..., m_t\}$ $\upsilon_{i,j} = 1 \ \forall i \in \{0, 1, ..., n\}, j \in \{0, 1, ..., m_t\}$
Repeat: $\upsilon_{i,j} = (\varpi_{0,j}\varsigma_{0,j} + \sum_{i'}\varpi_{i',j}\varsigma_{i',j})^{-(1-\gamma)} \times (\varpi_{0,j}y\varsigma_{0,j} + \sum_{i' \neq i}\varpi_{i',j}\varsigma_{i',j})^{-\gamma} \times e^k \ \forall i, j$ $\upsilon_{i,0} = (\varsigma_{i,0})^{1/\varepsilon - 1} \ \forall i$ $\upsilon_{0,j} = \left(\varpi_{0,j}\varsigma_{0,j} + \sum_i \varpi_{i',j}\varsigma_{i',j}\right)^{-1} \ \forall j$ $\varsigma_{i,j} = (\varpi_{i,0}\varsigma_{i,0} + \sum_{j'}\varpi_{i,j'}\varsigma_{i,j'})^{-(1-\gamma)} \times (\varpi_{i,0}\varsigma_{i,0} + \sum_{j' \neq j}\varpi_{i,j'}\varsigma_{i,j'})^{-\gamma} \times e^k \ \forall i, j$ $\varsigma_{i,0} = (\varpi_{i,0}\varsigma_{i,0} + \sum_{j'}\varpi_{i,j'}\varsigma_{i,j'})^{-1} \ \forall i$ $\varsigma_{0,j} = (\upsilon_{0,j})^{1/\beta - 1} \ \forall j$

(continued)

(continued)

The multi-hypothesis fractional belief propagation for single scan
Until: sufficiently small change in $\varsigma_{i,j}$, $\varsigma_{i,0}$, $\varsigma_{0,j}$
Calculate: $q_{0,0} \triangleq 0$ $q_{i,j} = \varpi_{i,j} \varsigma_{i,j} \upsilon_{0,j}$ $q_{i,0} = \varpi_{i,0} \cdot (\varsigma_{i,0})^{1/\varepsilon} \forall i$ $q_{0,j} = \varpi_{0,j} \cdot (\varsigma_{0,j})^{1/(1-\beta)} \forall j$
Output: $q_{i,j}$, $\forall i \in \{0, 1, ..., n\}$, $j \in \{0, 1, ..., m_t\}$

In each scan, MHFBP is trying to find the most suitable global hypothesis. After the optimal $q_{i,j}$ is obtained through the above algorithm, the meaningless measurement information can be eliminated in time, thus avoiding the exponential growth of the global assumption with the increases of target. At the same time, we update the previously assumed trajectory set (including trajectory update, trajectory termination, and new trajectory generation) to prepare for the next data association. While retaining all the hypotheses of the real goal, the hypotheses can be guaranteed to be passed on.

6.4.3 Experimental Design

Experimental scenario: An indoor environment with a length of 5 m and a width of 6 m.

Radar data acquisition: The data in this paper is obtained from the radar provided by Beijing TransMicrowave Technology Company. The system mainly includes radar, camera and antenna. The specific installation location of the radar is shown in the Fig. 6.38, and the relevant parameters of the radar are shown in Table 6.12.

Data preprocessing stage: In the actual tracking scene, the target detection data of the radar is a series of point traces. In order to correspond to the above assumption that

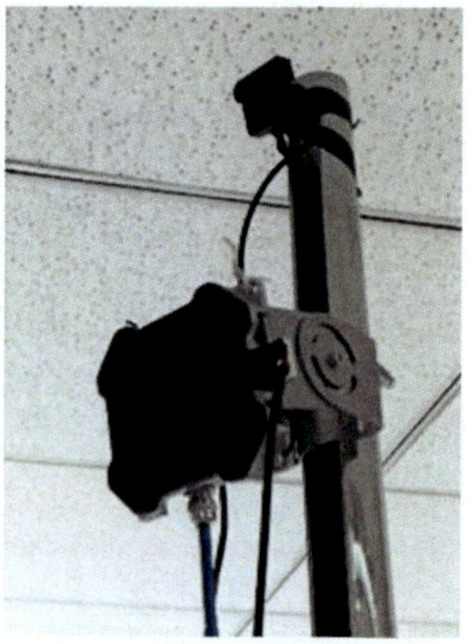

Fig. 6.38 The installation location of the radar

Table 6.12 The relevant parameters of the radar

Radar model	Radar system	The center frequency	Range resolution	Scan period	Scan range
CSR-EC-10	FMCW	77 GHz	10 cm	200 ms	FOV150°

each target generates at most one measurement data, the data needs to be clustered by Adaptive Possibilistic C-Means (APCM).

For one scan, from the actual scene, it can be clearly seen that in the radar detection range, in addition to two targets, it also includes obstacles such as tables and chairs. The echoes generated by these objects are all represented in the original data. According to the speed difference between these obstacles and moving targets, some obstacle points can be filtered out by setting a speed threshold (clutter cannot be completely filtered out). Just like the assumption made in the second part of the paper: each target is correlated with at most one measurement. Actually, the measurement data of each target in each scan is much more than one. Next, cluster analysis of the target is necessary. And we use the cluster center to represent the location information of the target. In this paper, APCM is used to cluster the targets, and each target cluster center is "o". Some of the clustering results in three application scenarios are shown in Fig. 6.39.

Data association: Here, we mainly reflect the performance of the algorithm proposed in this article by comparing it with the classic MHT, FA-MHT, and MHT-BP algorithms. MHT: As a classic data association algorithm, it has inspired many innovative data association algorithms. FA-MHT: Feature-Aided Tracking (FAT) has attracted attention due to its significant advantages over traditional target tracking. MHT-BP: Use Belief Propagation (BP) to approximate the marginal association probability (belief).

6.4.3.1 The Scenario of Two Targets

The results in Fig. 6.40 show the movement states of the two targets during the detection period. It should be noted that the initial movement times of the two targets are not same. It can be clearly seen from Fig. 6.41 that the data association result of the classic MHT is the worst, while the association performance of the other three algorithms is not significantly different. Therefore, here we use the position error (PE) between the estimated position of the target and the true ground position as a quantitative evaluation criterion to quantitatively compare the performance of these three algorithms. Here, because target 2 is a newly added target, the number of scan samples for target 1 and target 2 are different. In Fig. 6.42, by comparison, it can be found that the PE value of the proposed new algorithm is smaller than the other two algorithms. In order to test the data association ability of the new algorithm in a complex environment and obtain more objective conclusions, it is necessary to

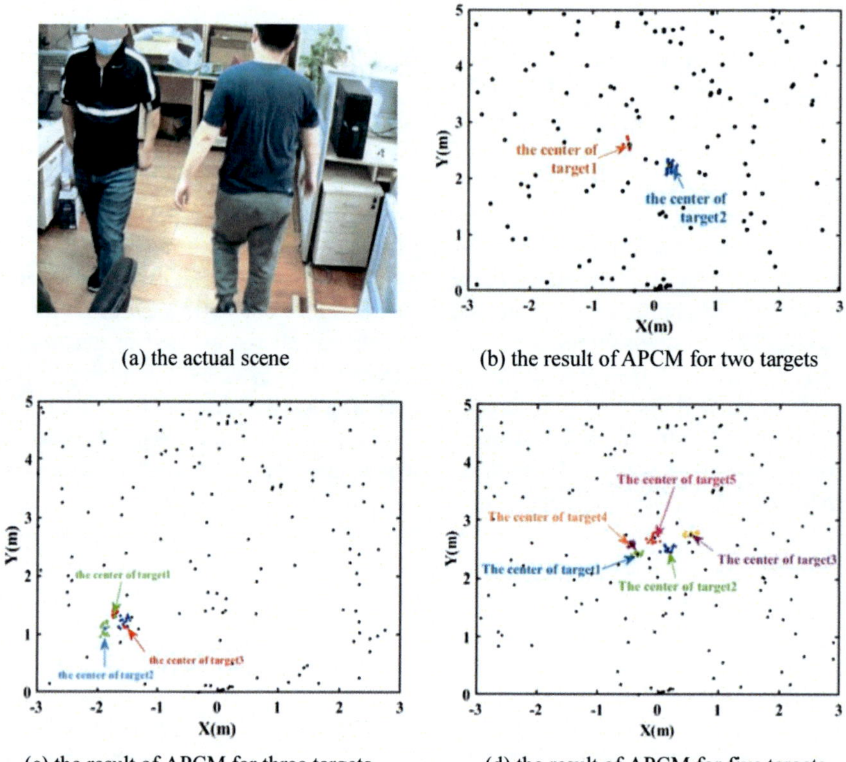

(a) the actual scene

(b) the result of APCM for two targets

(c) the result of APCM for three targets

(d) the result of APCM for five targets

Fig. 6.39 The results of APCM for the actual scenes

expand the number of the targets to 3, and increase the complexity of the tracking scene at the same time.

6.4.3.2 The Scenario of Three Targets

The motion state of these three targets is: Target 1 meets Target 3 and Target 2 respectively. After meeting with Target 2, the distance between the three targets is relatively close. As shown in Fig. 6.43, the result of MHT is still the worst. In particular, when the two targets walk closely and the trajectory crosses, in several consecutive scans MHT will have an error associated with false targets and some unfiltered clutter compared to other algorithms. Therefore, we need to use the root mean square error (RMSE) to compare the performance of the other three algorithms more objectively. Here, we conducted 50 simulation tracking experiments with three targets. In each experiment, we ensure that the tracks of at least two targets cross, or ensure at least two targets walk closely remaining at least 30 s. The results in Fig. 6.44 show the RMSE results of these three targets. The results in Table 6.13

Fig. 6.40 **a–d** represents the movement state of the two targets

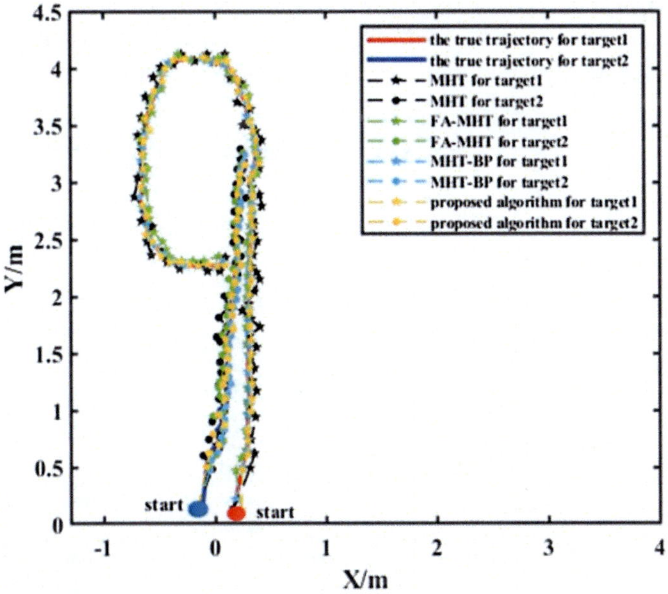

Fig. 6.41 The results of four different algorithms about two targets

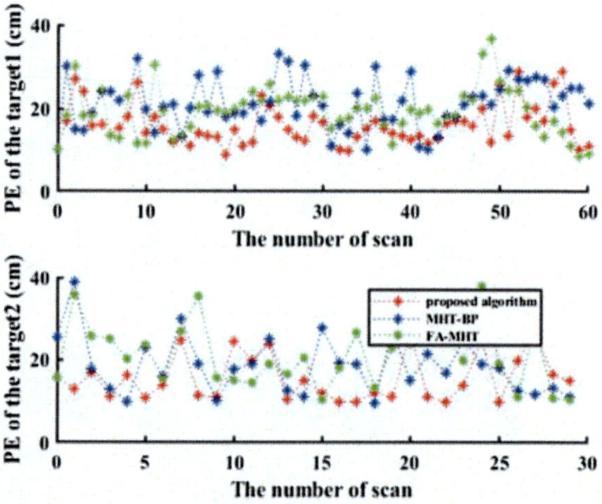

Fig. 6.42 The results of PE about three different algorithms

show the average value of the RMSE of these three algorithms. It can be seen that the algorithm proposed in this paper is 13.12% and 22.89% higher than the FA-MHT and MHT-BP algorithms respectively.

Similarly, as the complexity of the tracking scene increases, here we compare the tracking effects of the three algorithms by comparing their correlation accuracy rates in 60 s. The specific results are shown in Fig. 6.45.

It can be found from Fig. 6.45 that during this period of time, multiple-target encountering and multiple-trajectory crossing, the correlation accuracy of the two

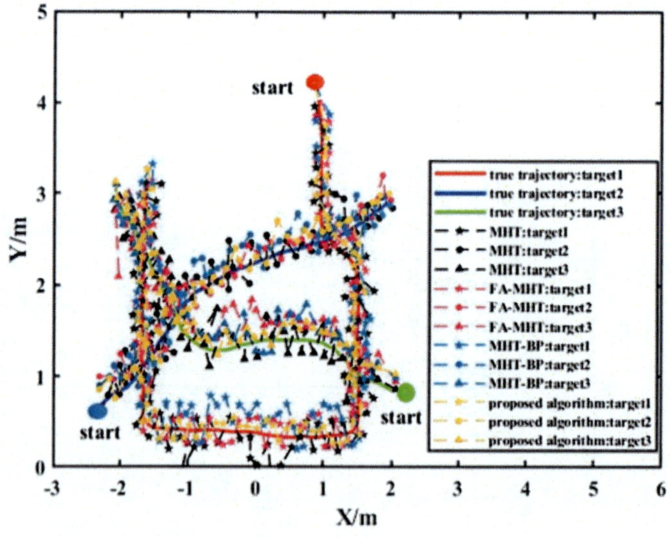

Fig. 6.43 The results of four different algorithms about three targets

Table 6.13 RMSE of proposed algorithm and other two algorithms for three targets

	Algorithm	Target 1	Target 2	Target 3	Improvement rate (%)
RMSE (cm)	FA-MHT	17.64	15.60	18.15	13.12
	MHT-BP	21.42	18.50	18.31	22.89
	Propose algorithm	14.26	14.20	16.11	

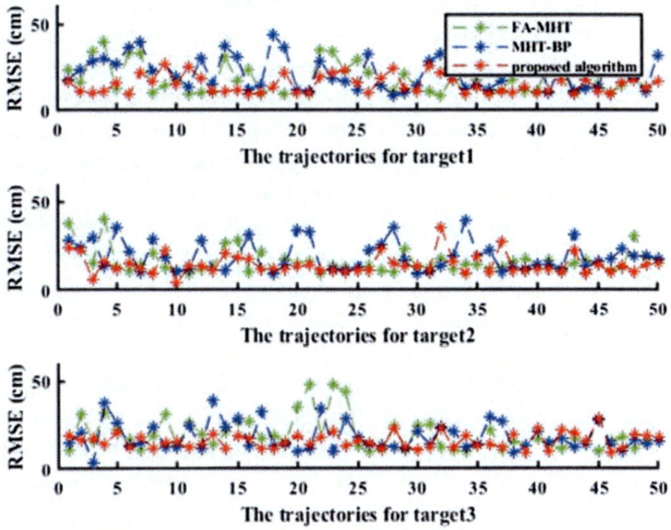

Fig. 6.44 The RMSE of three algorithms for three targets

Fig. 6.45 The correct rate of three algorithms

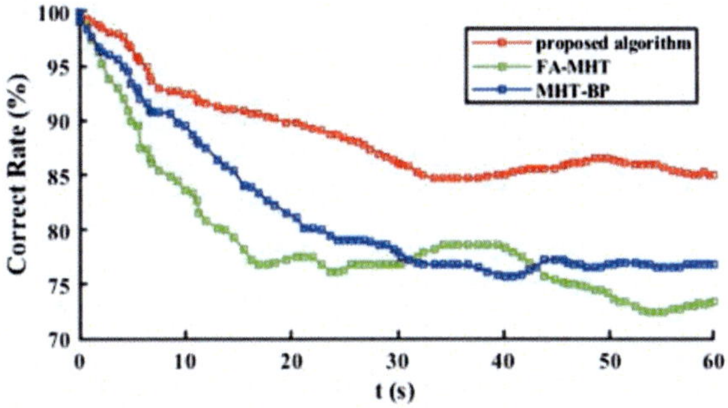

algorithms of FA-MHT and MHTBP quickly drops below 80%. In contrast, the speed change rate of the correlation accuracy of the new algorithm is small, indicating that MHFBP can estimate the marginal association probability distribution more accurately in the process of intensive multi-target tracking (IMTT). That's to say, the association belief generated by the proposed algorithm has a high certainty. With the gradual separation of multiple targets, the correlation accuracy of the three algorithms fluctuates slightly. However, the correct rate of FA-MHT and MHTBP is always lower than the proposed algorithm.

In addition, the results in Fig. 6.46 shows the cumulative distribution function (CDF) of the RMSE of the three algorithms in the above three scenarios. It can be found from the figure that at least 90% of the RMSE of the proposed algorithm are lower than 24.04 cm, which is significantly better than FAMHT and MHTBP algorithm. In summary, compared with classic MHT, FA-MHT and MHTBP, the new algorithm can maintain relatively accurate positioning and better correlation effect during intensive multi-target tracking. It should be noted that these improvements are based on sacrificing computational cost. Therefore, it may reduce computational efficiency. It is worth mentioning that in dense multi-target tracking scenarios, MHFBP can iteratively converge in a shorter time. Compared with the classic MHT, when the target and measurement increase, the algorithm can effectively avoid the trend of the exponential increase in the number of assumptions, and greatly reduce the computational complexity, fully reflecting the advantage of reducing the computational complexity from exponential to quadratic. Therefore, under the premise of ensuring the convex optimization of the objective function, the algorithm can solve the multi-objective data association problem while ensuring the accuracy of the association.

Moreover, various clustering algorithms with excellent performance are emerging in an endless stream. Although clustering algorithms are not our research content in this article, in the actual dense multi-target tracking scenario, the radar data needs to be speed filtered and clustered first. Although this operation will increase the overall running time of the program, it is also an essential key step in data association, and this two-stage MHT processing scheme can reduce the burden of the dynamic tracking system.

An MHFBP algorithm for the intensive multi-target tracking is proposed in this paper. The main work of this paper can be summarized as the following two points.

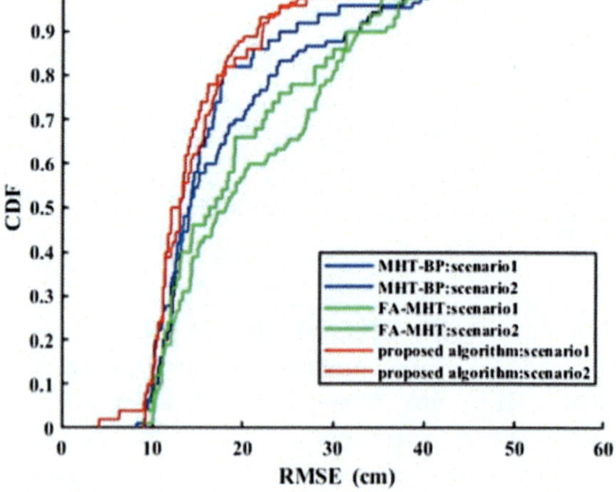

Fig. 6.46 The CDF of RMSE of three scenarios

(1) We apply variational inference to data association, establish a fractional free energy (FFE) function for track association indicator and track state variables, and optimize FFE by using convex optimization and FBP methods, and finally obtain reliable marginal association belief. The key of the new algorithm is to ensure the convexity of the objective function. In addition, the new algorithm will generate the same number of empty trajectories as the number of newly received measurements at each scan to take on these newly possible emerging targets, thereby effectively avoiding the targets missed.
(2) The new algorithm was applied into two, three, and five dense target tracking scenarios respectively, quantitatively analyzed from two aspects of positioning accuracy and correct rate, and compared with classic MHT and FA-MHT, MHT-BP. The comparison results in the three scenarios all show that the proposed algorithm has higher positioning accuracy and correlation accuracy than the other three algorithms. Therefore, the algorithm proposed in this paper has the potential to be applied in more complex tracking scenarios. In the next work, we consider applying this algorithm to more challenging outdoor tracking scenes containing more targets.

6.5 Design of Multi-target Tracking System

The purpose of multi-target tracking system design is to realize real-time tracking of targets through software. The main functions of the software include dynamic display of the real situation of the scene, adjustment of radar-related parameters, and display of real-time vehicle data.

6.5.1 Requirements Analysis

The multi-target tracking system involved in this section relies on the CSR-TS radar independently developed by Beijing Chuansu Microwave Technology Co., Ltd. This product adopts the international advanced microwave radar technology, and has made important breakthroughs in radar system, signal processing, target recognition algorithm and so on. Traffic surveillance radar can be used for high-speed road condition monitoring and intelligent control of traffic lights, tracking the driving trajectories of all vehicles in the current surveillance scene, and obtaining information such as the location and speed of each vehicle. The radar can not only provide traditional microwave vehicle detection information such as traffic flow, lane occupancy, and average vehicle speed, but also provide information such as wrong-way vehicles, emergency-stop vehicles, lane-changing vehicles, and queue lengths. Traffic surveillance radar will become an indispensable front-end equipment in smart traffic scenarios.

Fig. 6.47 Application scenario of traffic surveillance radar

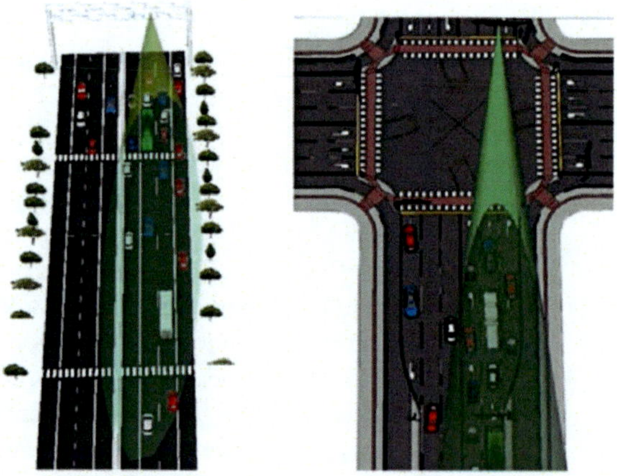

(a) Scene ① (b) Scene ②

CSR-TS radar is widely used in high-speed road condition monitoring and traffic light intelligent control. There are two main application scenarios: (1) highway and urban highway road condition detection, and alarm for wrong-way and illegal lane change; (2) provide queue length information for intelligent control of traffic lights at intersections. The application scenario of traffic surveillance radar is shown in Fig. 6.47.

The main functions of the CSR-TS radar are as follows.

(1) Real-time tracking of multiple targets.

① It can monitor 4 to 5 lanes at the same time, and the number of monitoring lanes can be adjusted.
② It has high speed and ranging resolution.
③ It can know the speed and direction of the vehicle.
④ The trajectory tracking accuracy rate exceeds 90%.

(2) Statistics of traffic flow information.

① Using the statistical function of the supporting software to realize traffic flow statistics.
② Count the number of passing vehicles in each lane.
③ Count the number of passing vehicles in each lane in unit time.
④ Calculate the average speed of each lane and the average speed of all lanes.

(3) Incident monitoring.

① Real-time monitoring and judgment of wrong-way vehicles.
② Real-time monitoring and judgment of lane-changing vehicles.
③ Real-time monitoring and judgment of emergency stop vehicles.
④ Determine the status of queues in traffic jams.

Figure 6.48 shows the installation diagram of the CSR-TS radar. The installation height is 6 to 10 m. The width of the radar beam coverage area is related to the beam center position.

CSR-TS radar is mainly used for trajectory tracking and event monitoring of vehicles in multiple lanes. Therefore, the radar is usually fixed at the center position in the width direction, that is, the center position of the gantry above the road.

In order to narrow the application scope of radar monitoring, the installation requirements can be appropriately reduced, and the actual installation position of the radar can deviate from the center position of the total lane width. Therefore, in addition to the top-mounted, the radar also has side-mounted, side-mounted and other installation methods.

(1) Top-mounted

The radar is installed just above the monitoring lane and is suitable for the situation with a gantry. Figure 6.49 shows the schematic diagram of the top-mounted radar.

(2) Side top mounted

The radar is installed above the monitoring lane and deviates from the center of the lane at any distance, which is suitable for the situation with L-bar. Figure 6.50 shows the schematic diagram of the radar side top installation.

(3) Side mounted

The radar is installed on the side of the road, at a distance from the outermost lane L (0–3 m), which is suitable for the situation with poles. Figure 6.51 shows the schematic diagram of radar side installation.

Fig. 6.48 The schematic diagram of CSR-TS radar installation

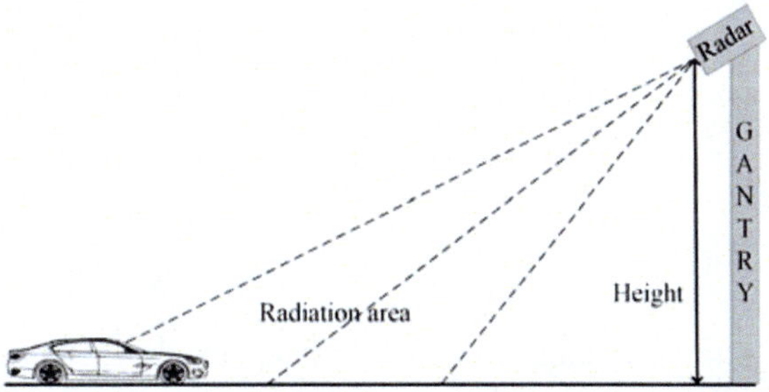

Fig. 6.49 The schematic diagram of the top-mounted radar

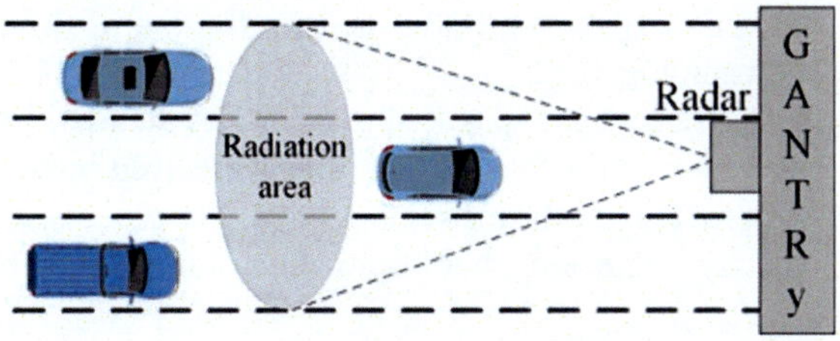

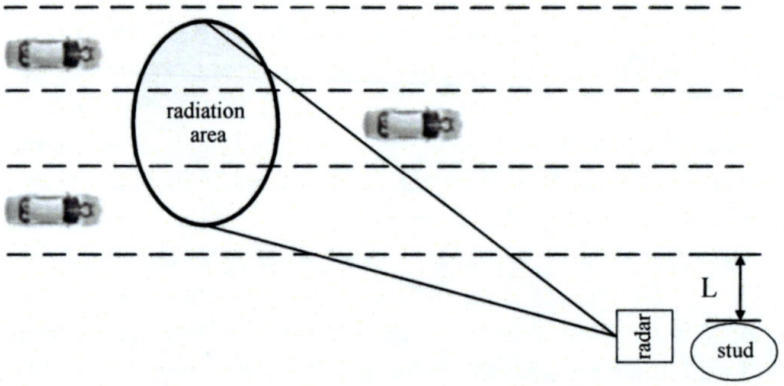

Fig. 6.50 The schematic diagram of radar side top installation

Fig. 6.51 The schematic diagram of radar side installation

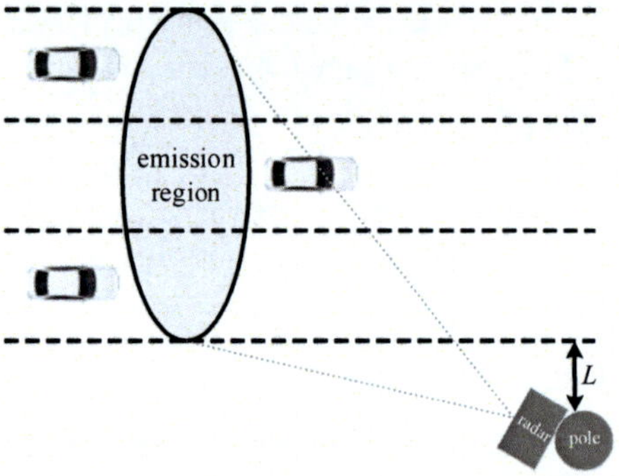

In order to better adapt to the CSR-TS radar, the designed multi-target tracking system needs to support all the usage scenarios, main functions and installation methods of the CSR-TS radar. Based on this requirement, the overall design of the multi-target tracking system is carried out.

6.5.2 Overall Design

The multi-target tracking system mainly includes three modules: radar signal processing, radar data processing and processing result display. Radar data processing algorithms include clustering algorithms, data association algorithms, and filtering prediction algorithms. The module division of the multi-target tracking system is shown in Fig. 6.52.

The radar signal processing module is the first running module in the system. Its function is to process the echo signal received by the radar, obtain the required data and process the data. In this section, the processed data includes the coordinates, velocity, angle and other information of the sample points.

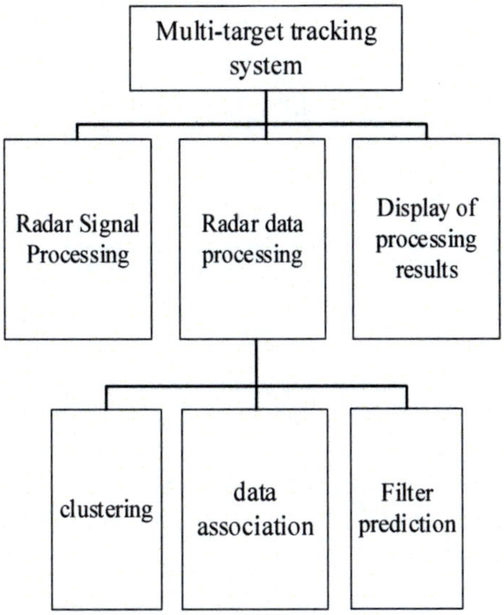

Fig. 6.52 Module division of multi-target tracking system

The radar data processing module is the most critical module in the system, and its function is to reprocess the data processed by the radar signal processing module. This module uses clustering algorithm, data association algorithm and filter prediction algorithm to track all vehicles in the scene.

The processing result display module is the module used to present the results in the system. Through the previous signal processing and data processing, multi-target tracking can be completed. In addition, visualization is an important step in showing the operating status of the system. In this module, the complete multi-target tracking process can be displayed through the system interface design and the visualization of the operation results.

The multi-target tracking system process can be simply divided into human–computer interaction process, interface process and processing process. The human–computer interaction process is the process in which the user uses the computer to operate the software to complete operations such as status monitoring, equipment management, and parameter setting. The interface process is the bridge between the human–computer interaction process and the processing process. The processing process includes the radar signal processing module and the radar data processing module. The process division of the multi-target tracking system is shown in Fig. 6.53.

The radar signal processing module completes the conversion between the radar echo signal and the processable data. The steps are as follows.

(1) Receive radar echo signals.
(2) Transform the echo signal from the time domain to the frequency domain.

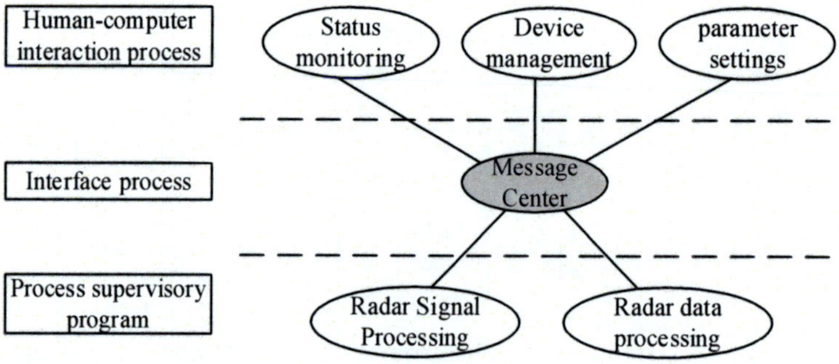

Fig. 6.53 Multi-target tracking system process division

(3) Calculate the distance between the target and the radar.
(4) Calculate the angle between the target and the radar.
(5) Calculate the coordinates of the target in the radar coordinate system.

The radar signal processing flow is shown in Fig. 6.54.

The radar data processing module needs to use the clustering algorithm to cluster the data processed by the radar signal processing module, and perform data association. At the same time, the filtering prediction algorithm needs to cooperate with the first two algorithms to cyclically process the signal of each frame, and finally obtain the multi-target motion state and complete the tracking process. The radar data processing flow is shown in Fig. 6.55.

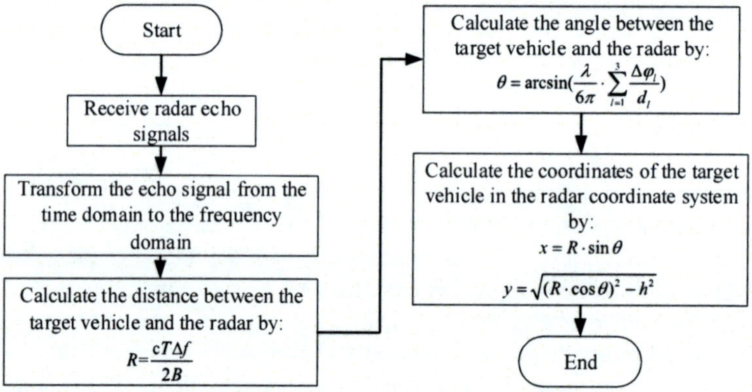

Fig. 6.54 Radar signal processing flow

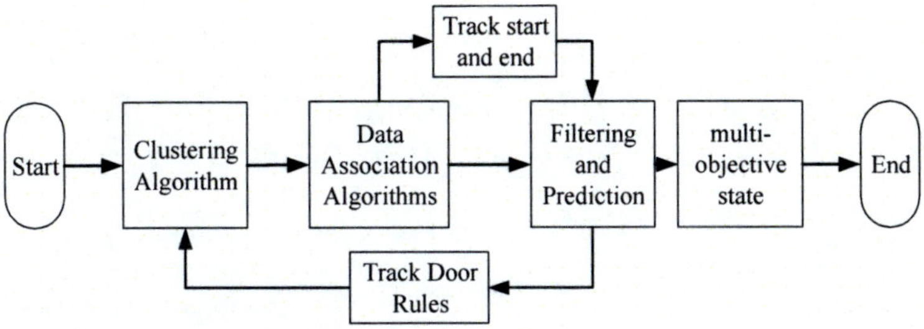

Fig. 6.55 The flow of Radar data processing

6.5.3 Application of Multi-target Tracking System

This section realizes the debugging and visualization of the multi-target tracking system on the computer through software. This software is the supporting software of the millimeter-wave surveillance radar. Its main functions include dynamic display of real-time road conditions, free debugging of system parameters, and real-time update of the target list.

1. Operating environment and interface display

This software can run on Windows 7 system and is compatible with Windows 10 system. It can dynamically display the driving conditions of all vehicles in the specified road section in real time, so that users can understand the driving status of the vehicles at the first time. The statistical data function of the software can display the traffic flow, queue length and other information of the specified road section. The software supports the query and setting of radar and lane-related parameters, and can display real-time vehicle data and statistical data. The main interface of the software is shown in Fig. 6.56.

The main interface of the software includes the following functional areas.

(1) Title bar: Display the software name and version, as shown in Fig. 6.57.
(2) Real-time road conditions: Dynamically display the real-time road conditions in the monitored road section, as shown in Fig. 6.58.
(3) Device management: The communication between the software and the radar can be established by modifying the relevant network parameters. The device management module is shown in Fig. 6.59.
(4) Parameter query: Query the parameters related to radar and lane, as shown in Fig. 6.60.
(5) Parameter setting: Set the parameters of radar and lane, as shown in Fig. 6.61.
(6) Road condition data: Refresh the list of real-time data or statistical data, as shown in Fig. 6.62.
(7) Status bar: Display the working status of the system, as shown in Fig. 6.63.

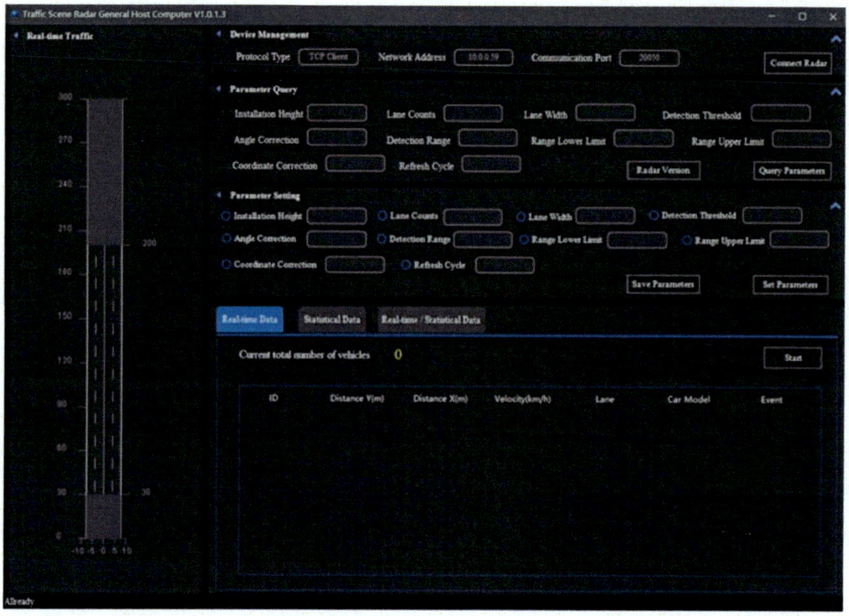

Fig. 6.56 Main interface of the software

Fig. 6.57 Title bar

2. Operation steps

The default values in device management and parameter settings come from the configuration file AppConfig.xml, as shown in Fig. 6.64. The corresponding relationship between the parameters in the configuration file and the parameters of the main interface of the software is as follows.

The corresponding relationship of parameters	
Device management	
Configuration file	Main interface of the software
Ip	Network address
port	Communication port
Parameter setting	
radar_height	Installation height

(continued)

(continued)

lane_number	Number of lanes
lane_width	Llane width
crc_threshold	Detection threshold
angular	Angle correction
detection_threshold	Detection range
ymin	The lower limit of ranging
ymax	The upper limit of ranging
coordinate	Coordinate correction
refresh_time	Refresh period
lane_line_type	The lane line type of 12 lanes can be set

Lane_line_type is a maximum of 11 digits, corresponding to the right lane lines of 11 lanes, each digit is represented by 0 or 1, 0 means the lane line is a dashed line, and 1 means a solid line.

In order to establish the communication between the multi-target tracking software and the radar equipment through the network, it is necessary to set the network address and communication port first. After filling in the network address and communication port of the radar to be connected correctly, click "Connect Radar", if the connection is successful, the status bar will display "Connected…successful".

The parameters of the network address and communication port are obtained from the configuration file, which can be modified through the text box in the main interface of the software. If you reopen the software, the parameters of the network address and communication port default to the values set during the last communication.

In the parameter query module, click "Query Parameters", and the radar parameters of the lower computer will be displayed in the corresponding text box; click "Radar Version", the status bar will display the current radar firmware version.

The installation height in the parameter setting module is the vertical height of the radar from the ground. The angle correction can correct the angle between the radar and the road. The number of lanes and lane width should be set according to the actual situation. The farthest detection distance of the radar is 250 m, but the "lower limit of ranging" and "upper limit of ranging" can be adjusted according to the actual monitoring needs.

After the software establishes a connection with the radar device, the current road condition data can be displayed by clicking "Start" in the road condition data module. The road condition data includes real-time data and statistical data. The real-time data counts the current number of vehicles on the road section, and displays the status information of all vehicles in the road section in tabular form. Statistical data displays information such as traffic flow and queue length in the current road section.

3. Practical application

Click "Start" in the road condition data to display and refresh the vehicle data in real time.

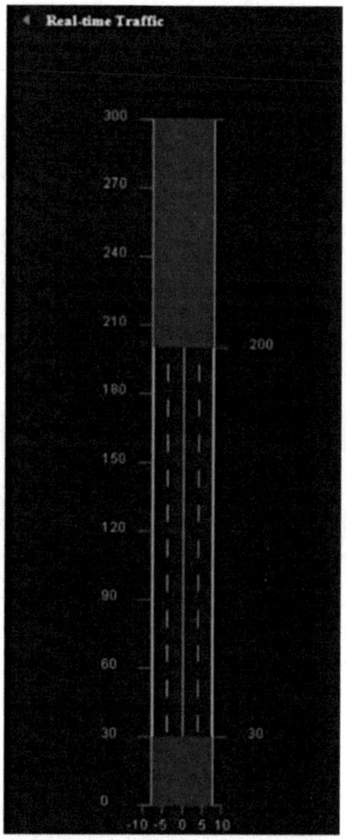

Fig. 6.58 Real-time road conditions

Fig. 6.59 Device management module

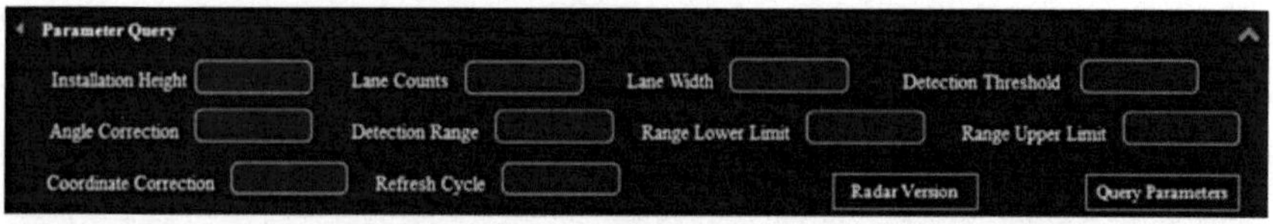

Fig. 6.60 Parameter query module

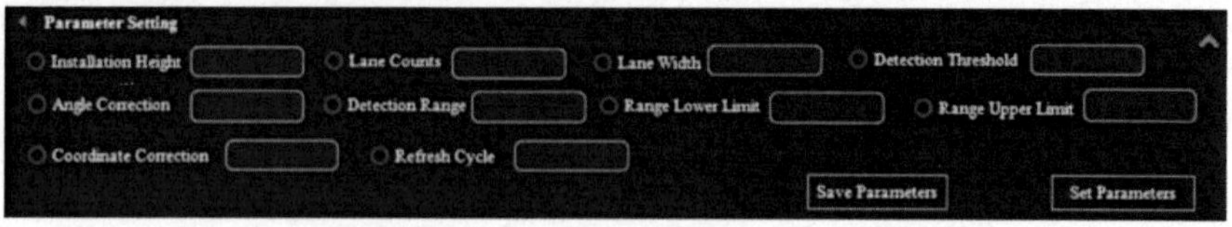

Fig. 6.61 Parameter setting module

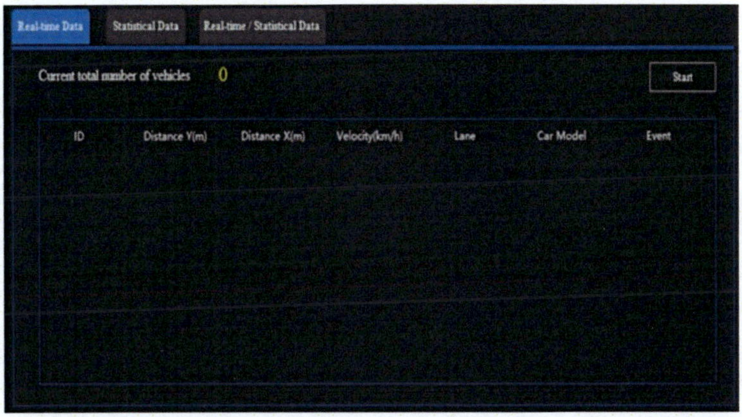

Fig. 6.62 Road condition data display module

Fig. 6.63 Status bar

```
<?xml version='1.0' encoding='UTF-8'?>
<config>
    <tcp>
        <ip>192.168.100.7</ip>
        <port>6172</port>
    </tcp>
    <parameter>
        <radar_height>7.00</radar_height>
        <lane_number>4</lane_number>
        <lane_width>3.75</lane_width>
        <crc_threshold>2000</crc_threshold>
        <angular>0.00</angular>
        <detection_threshold>2</detection_threshold>
        <ymin>30</ymin>
        <ymax>200</ymax>
        <coordinate>0</coordinate>
        <refresh_time>1</refresh_time>
    </parameter>
    <ui>
        <lane_line_type>110111111111</lane_line_type>
    </ui>
</config>
```

Fig. 6.64 Configuration file

Fig. 6.65 Program running interface

The program running interface is shown in Fig. 6.65. From Fig. 6.65, you can clearly see the vehicle's position, speed, lane and other information.

As can be seen from Fig. 6.65, there are 4 large vehicles on the monitoring road, 3 of which are far away from the radar, and 1 vehicle has just appeared in the field of vision.

It can be seen from the above data that due to the influence of the radar installation angle and parameters, the targets closer to the horizontal axis of the radar cannot be displayed on the software. The target needs to have a certain distance from the radar to be detected by the radar. Multiple and long-term engineering practices have verified that the designed multi-target tracking system can be used in conjunction with millimeter-wave surveillance radar in actual traffic scenarios, and can maintain high accuracy and good real-time performance in the multi-target tracking process. The program running interface in some scenarios is shown in Fig. 6.66.

6.6 Summary

This chapter divides the track association into two simple steps, adds constraints according to the radar configuration, completes the track association, and combines the Kalman filter and the particle filter to propose the PF-KIS algorithm. The drive test results show that the tracking method used in this chapter can meet the real-time requirements, improve the accuracy of nonlinear estimation, and has a good multi-target tracking effect. In order to track multiple targets more accurately, this chapter proposes the k-Nearest Neighbor Joint Probabilistic Data Association (kNN-JPDA) algorithm, and discusses the advantages of the data association algorithm and its importance in multi-target tracking systems. This chapter also introduces some classic data association algorithms (NNDA and JPDA) and compares them with the kNN-JPDA algorithm. The kNN-JPDA algorithm proposes a new pre-aggregation

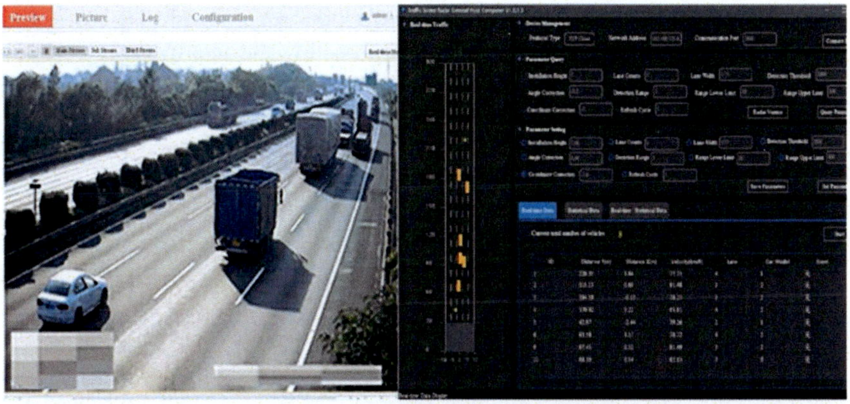

(a)

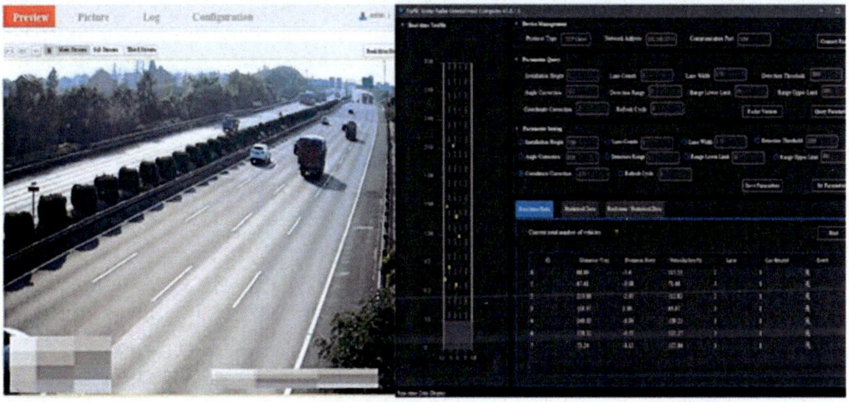

(b)

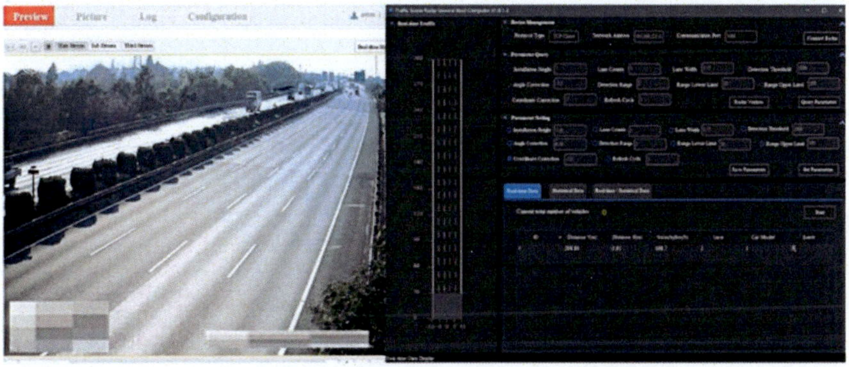

(c)

Fig. 6.66 Program running interface in some scenarios

matrix and aggregation matrix, which improves the accuracy of data association by calculating the aggregation correlation probability, and improves the operation efficiency of the algorithm by constructing the aggregation matrix. In this chapter, it is proved by experiments that the kNN-JPDA algorithm has high accuracy and efficiency. In the actual road test, the algorithm can also complete the multi-target data association well, so as to achieve multi-target tracking.

In order to solve the multi-target tracking problem in traffic scenes, this chapter integrates the improved algorithms and designs a multi-target tracking system. In this system, users can intuitively see the process of receiving echo signals through the millimeter-wave radar and completing the multi-target tracking task in real time. This chapter clarifies the design requirements of the multi-target tracking system, and introduces the design idea, operation process and usage method of the multi-target tracking system. Finally, multiple tests are carried out in real traffic scenarios to verify the feasibility and stability of the proposed multi-target tracking system.

Chapter 7
Information Fusion and Target Recognition Based on Millimeter Wave Radar and Machine Vision

In the car assisted driving system, target recognition is a difficult point in the application of low-resolution radar. It only uses one type of sensor as a data source, which has certain defects in reliability. Accurate information can only be obtained by fusing the information measured by multiple sensors. Vehicle detection can be performed using images captured by cameras and echo data captured by millimeter-wave radars. In terms of details, compared with millimeter-wave radar, cameras have obvious advantages, but as the frequency of millimeter-wave radar increases, its resolution increases rapidly, and it can also complete some relatively fine work.

This chapter uses deep learning algorithms to detect objects from images and give their pixel locations. Convolutional Neural Network (CNN) is the first deep learning algorithm to achieve success in practice. CNN is not generated out of thin air. It is built on the basis of classic feedforward neural network and draws on filtering and edge detection in the traditional image field. Compared with traditional image processing algorithms, CNNs pay more attention to data, so that the model automatically learns features from the data without the need for human-designed features. For the classical artificial neural network, the input dimension of the image is too high, and when training the model, a huge data set and hardware memory are required. In order to solve these problems, the CNN introduced convolution and reduced the number of network parameters from hundreds of millions to millions through "parameter sharing" and "sparse connection", and achieved great success in the field of images, from the effect surpasses traditional algorithms.

L. Cao et al., *Target Recognition and Tracking for Millimeter Wave Radar in Intelligent Transportation*, https://doi.org/10.1007/978-981-99-1533-0_7

7.1 Target Detection Based on Deep Learning

7.1.1 Target Detection

1. ResNet

In this chapter, the data fusion of millimeter-wave radar and camera is realized. Based on the data fusion model of millimeter-wave radar and camera, the coordinate transformation relationship between radar and camera is established, and the fusion of radar data and camera data in space and time dimensions is realized. In addition, according to the fusion results, the conversion from image data positioning to radar data positioning can be realized. According to the target recognition results of the image, the target category labels are provided for the radar data, and feature extraction is performed based on the characteristics of the radar data, using machine learning algorithms. Train and classify labeled radar data for target recognition.

Theoretically speaking, the deeper the layer of the neural network, the more complex the mapping that the model can learn. However, as the depth of the network increases, there are phenomena of gradient disappearance and gradient explosion. In order to solve this problem, He Kaiming proposed a new network structure: Residual Network (ResNet). Compared with the traditional network structure, ResNet adds a "shortcut" to the network connection, called "skip connection", as shown in Fig. 7.1. For the traditional network, let the output of the Lth layer of the model be $a^{[L]}$, the parameters are $W^{[L]}$ and $b^{[L]}$, and the excitation function is relu, then a is

$$a^{[L+1]} = \text{relu}(W^{[L]}a^{[L]} + b^{[L]}) \tag{7.1}$$

After adding "Skip Connection", $a^{[L+1]}$ is

$$a^{[L+1]} = \text{relu}(W^{[L]}a^{[L]} + b^{[L]}+a^{[L]}) \tag{7.2}$$

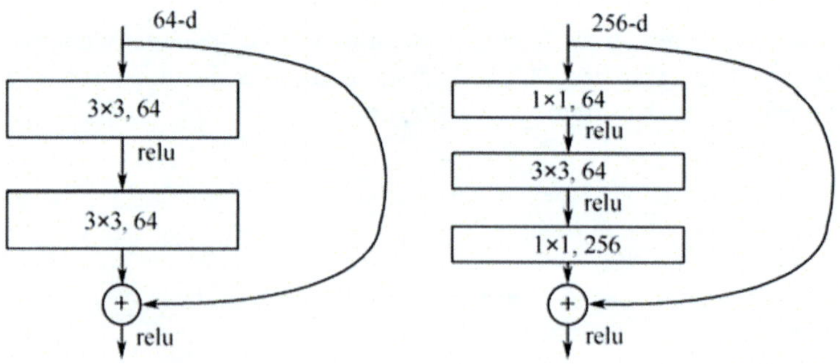

Fig. 7.1 Skip connection

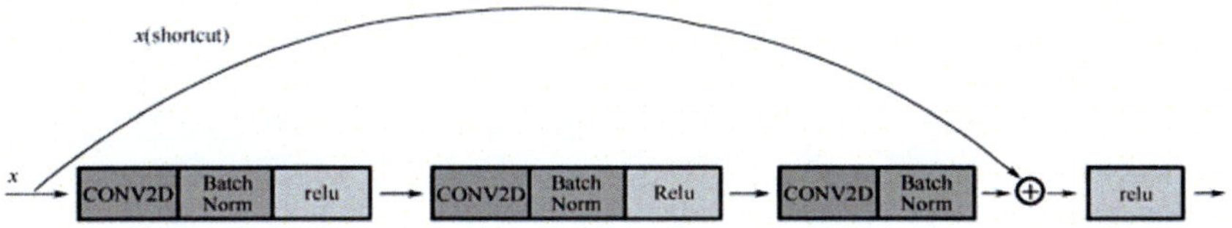

Fig. 7.2 Identity block

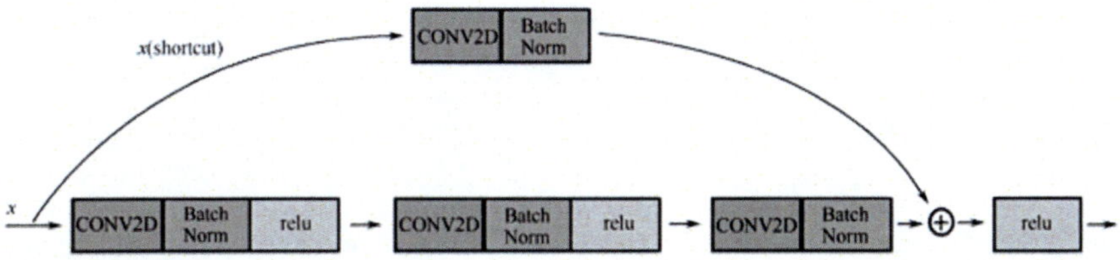

Fig. 7.3 Convolutional block

When the gradient disappearance problem occurs, the parameters $W^{[L]}$ and $b^{[L]}$ become very small and are approximately 0. At this time, the output of this layer is $a^{[L+1]} = \text{relu}(a^{[L]}) = a^{[L]}$, , which means that the learning of this layer will not reduce the excitation of the previous layer at least (not will decay to 0), so using "skip connection" can effectively train deep networks.

In addition to the above problems, we also need to consider the matrix dimension matching problem. If the dimensions of the matrix obtained through "skip connection" match with the dimensions of the network model matrix, the two are directly added to get Identity Block, as shown in Fig. 7.2; if the dimensions of the matrix do not match, a convolution layer is added on the path of "skip connection", so that the dimensions of the two can match normally, thus ensuring that the multiplication and convolution operations of the matrix will not go wrong. This structure is called Convolutional Block, as shown in Fig. 7.3.

2. InceptionNet

In the convolutional network structure, the size of the convolution kernel affects the quality of the extracted features and the performance of the algorithm. The convolution kernel used in the early convolutional network structures (such as LeNet-5 and AlexNet) is larger (5×5), because the higher the dimension of the convolution operator, the larger the corresponding receptive field, and the convolution kernel. More information is seen in the image, and therefore better features are obtained. However, it can be found from VGG-16 that the effect of using two 3×3 convolution kernels is better than using one 5×5 convolution kernel, and can significantly reduce the amount of calculation, so the 3×3 convolution kernel is currently the most widely used.

Is it true that each convolutional layer can only use convolutional kernels of the same size?In 2014, the InceptionNet structure proposed by Google first appeared in

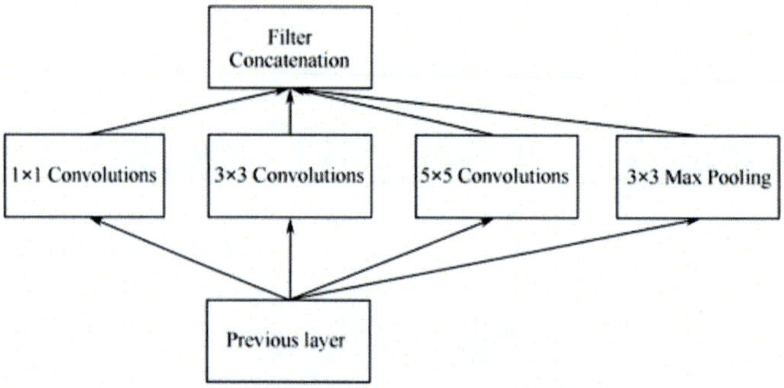

Fig. 7.4 InceptionNet

the ILSVRC competition and won. InceptionNet convolves the output of the previous layer with multiple convolutional kernels of different sizes to obtain features at different scales. InceptionNet is shown in Fig. 7.4. The output of the previous layer is processed by 1 × 1, 3 × 3 and 5 × 5 convolution kernels and 3 × 3 pooling, and these features are combined to obtain the output of the current layer.

The disadvantages of inception net are large amount of calculation and low efficiency. In order to solve this problem, the Google team was inspired by the structure of NIN (network in network) by adding 3 × 3 convolution kernel (called bottleneck layer) significantly reduces the amount of calculation and increases the inception net of bottleneck layer, as shown in Fig. 7.5.

Assuming that the input dimension are 28 × 28 × 192, using 32 convolution cores with 5 × 5 × 192 dimensions needs to complete $28 \times 28 \times 16 \times 192 + 28 \times 28 \times 32 \times 5 \times 5 \times 16 \approx 0.12$ million multiplication operations. After using Bottleneck layer, you can first use 6 convolution cores with 1 × 1 × 192 dimensions for processing, and then use 32 convolution cores with 5 × 5 × 16 dimensions for processing. You need to complete $28 \times 28 \times 16 \times 192 + 28 \times 28 \times 32 \times 5 \times 5 \times 16 \approx 0.12$ billion

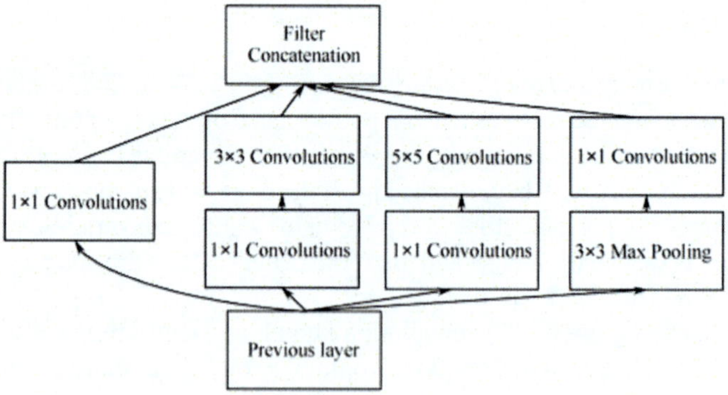

Fig. 7.5 Added the ConceptNet of bottleneck layer

multiplication operations, and the amount of calculation is about one tenth of the original!

In summary, both InceptionNet and ResNet are good structures and are used in target detection algorithms and various large-scale deep learning networks.

7.1.2 Principle of Deep Learning Target Detection

At present, target detection based on convolutional neural network can be divided into two categories: R-CNN series algorithm based on candidate region and YOLO algorithm based on grid division. The core idea of the candidate region based algorithm represented by R-CNN series algorithms is to use classification to solve the problem of target location. Specifically, first obtain the region where there may be targets through clustering algorithm, and then judge each candidate region by mature classification algorithm. The advantage is that it makes full use of the existing mature classification algorithms, and the accuracy of target detection is high. The disadvantage is that the efficiency of the algorithm is low, and most of the time is spent in the extraction process of candidate regions.

The core idea of the mesh based algorithm represented by Yolo is to regard the target detection task as a regression task. In this regression task, it is necessary to predict the pixel position and confidence probability of the target in the image, mark the label corresponding to the target as $y = (p_c, b_x, b_y, b_h, b_w, c)$, p_c represents the confidence probability of the target in the image, and b_x, b_y, b_h, b_w represents the abscissa, ordinate, height and width of the target respectively, The information of these four dimensions is called the Bounding box of the target, as shown in Fig. 7.6. c indicates the category of targets. When only vehicles are detected, c is a scalar; When it is necessary to detect multiple types of targets, c is usually a one- hot vector.

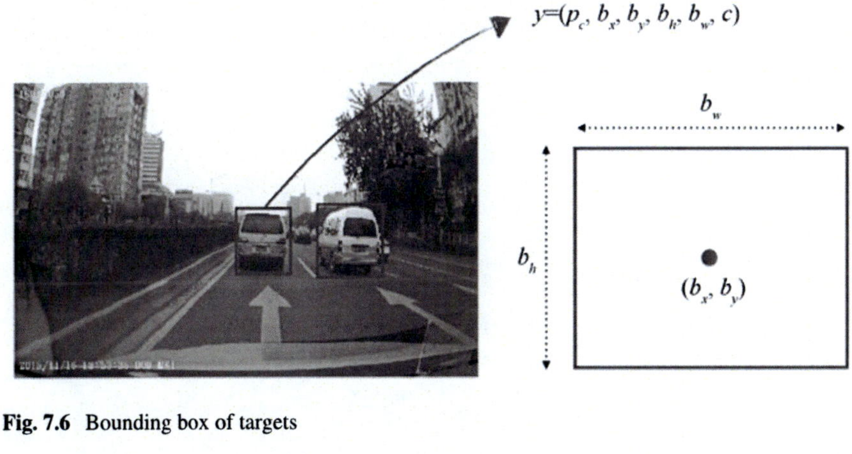

Fig. 7.6 Bounding box of targets

If there are multiple targets in the image, there will be multiple labels, that is, a sample will have multiple correct answers. If the traditional method is adopted, it will lead to failure of training. In order to solve this problem, YOLO algorithm normalizes the image to the size of 224 × 224 × 3 and divides the image into grids of $s \times s$. each grid is regarded as a subgraph, and each subgraph has a label. If the subgraph contains a target, then $p_c = 1$, otherwise $p_c = 0$. The grid division of YOLO algorithm is shown in Fig. 7.7. Figure 7.7 is divided into 3 × 3 grids, so that c is a three-dimensional vector, indicating that it is hoped to identify three types of targets from the image, namely people, vehicles and background. Then the label y corresponding to the sample is shown in Eq. (7.3), the label component y_i corresponds to the i th grid, and the '?' indicates the value we don't care about. In practical application, some small random numbers can be generated.

$$y = \begin{bmatrix} y_1 & y_2 & y_3 & y_4 & y_5 & y_6 & y_7 & y_8 & y_9 \\ 0 & 0 & 0 & 0 & 1 & 1 & 0 & 0 & 0 \\ ? & ? & ? & ? & b_x & b_x & ? & ? & ? \\ ? & ? & ? & ? & b_y & b_y & ? & ? & ? \\ ? & ? & ? & ? & b_h & b_h & ? & ? & ? \\ ? & ? & ? & ? & b_w & b_w & ? & ? & ? \\ ? & ? & ? & ? & 0 & 0 & ? & ? & ? \\ ? & ? & ? & ? & 1 & 1 & ? & ? & ? \\ ? & ? & ? & ? & 0 & 0 & ? & ? & ? \end{bmatrix} \tag{7.3}$$

Fig. 7.7 Mesh generation of YOLO algorithm

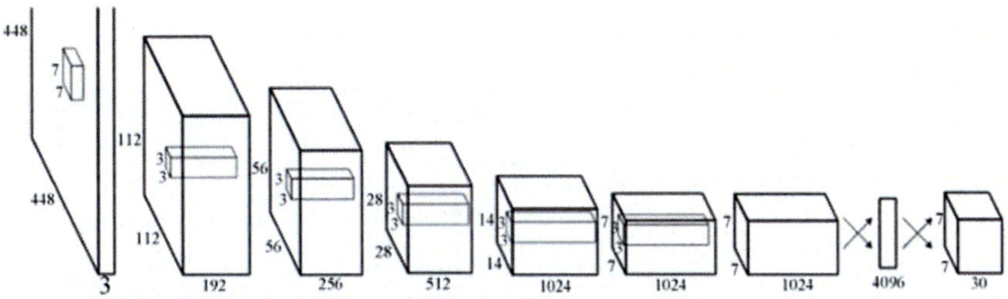

Fig. 7.8 The network structure of YOLO algorithm

To sum up, compared with traditional target detection algorithms (such as sliding window algorithm), YOLO algorithm can learn Bounding boxes of any size through regression, while sliding window algorithm can only give Bounding boxes of specified size. Compared with R-CNN series algorithms, YOLO algorithm realizes one-step detection, that is, Yolo algorithm is an end-to-end learning system, which is much faster than traditional algorithms and R-CNN series algorithms. It can reach 45fps on Titan X GPU, which can meet the needs of real-time detection.

YOLO algorithm not only greatly improves the efficiency, but also reduces the accuracy of target detection. In order to solve the problem of insufficient accuracy and easy loss of small targets, YOLOv2 algorithm has been improved from many aspects on the basis of the original algorithm. This section analyzes the three main improvements, introduces the training label based on Anchor box and the process of target detection based on the output results of convolution network by using non maximum suppression algorithm.

1. The core idea of YOLOv2 algorithm

The network structure of YOLO algorithm is shown in Fig. 7.8. The improvement of yolov2 algorithm based on the original algorithm is analyzed from three aspects.

Firstly, when designing the convolution network structure, YOLOv2 adds a BN layer behind each convolution layer and removes the dropout layer. BN layer is proposed to solve the "covariate shift" problem. In short, "covariate shift" refers to the problem that the training model will fail when the distribution of test set data and training set data is different. For neural networks, due to the strong coupling between layers, the algorithm itself has a serious "covariate shift" problem, resulting in the slow training speed. BN layer can normalize the input data into data with mean value of 0 and variance of 1, and add parameters γ and β that need to be learned through back propagation. If the mean value of data set $Z = \{z^{(1)}, z^{(2)}, \cdots, z^{(m)}\}$ is μ and the variance is σ^2, the output $\tilde{z}^{(i)}$ of $z^{(i)}$ after passing through BN layer can be expressed as

$$z_{\text{norm}}^{(i)} = \frac{z^{(i)} - \mu}{\sqrt{\sigma^2}} \tag{7.4}$$

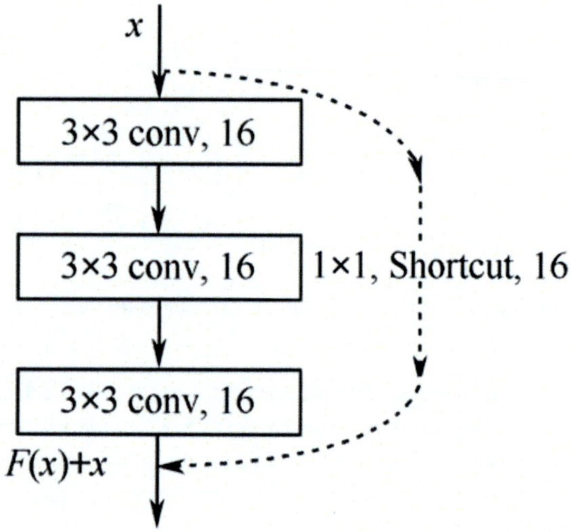

Fig. 7.9 Schematic diagram of the residual module in DarkNet-19

$$\tilde{z}^{(i)} = \gamma z_{\text{norm}}^{(i)} + \beta \tag{7.5}$$

Secondly, YOLOv2 algorithm adopts the 19 layer new network structure darknet-19 as the pre training network. The residual module is introduced into the network structure, as shown in Fig. 7.9, which can solve the problem of small target recognition to a certain extent. The "skip connection" structure in the residual module can combine the fine features learned in the first few layers with the advanced features learned in the last few layers. This idea is also reflected in SSD algorithm.

Finally, in order to make the algorithm have better detection effect on images with different resolutions, YOLOv2 algorithm adopts full convolution network structure, that is, 1 × 1 instead of the original full connection layer, so that the input of the convolution network does not have to be limited to a fixed size. Specifically, during network training, the size can be changed every 10 epoches for training.

2. Anchor box

The generation process of training labels in the grid-based algorithm was introduced earlier. When a target may appear in different grids, it can be considered that the grid where the center pixel of the target is located is the grid where the target is located, but If there are multiple targets in a grid, the training label can only give the information of one target. In order to solve this problem, Anchorbox is introduced.

Anchor box defines multiple bounding boxes with different shapes in advance for each mesh. Anchor box is shown in Fig. 7.10. For any bounding box, it can always be adjusted to three new bounding boxes according to the aspect ratio of 1:1, 1:2 and 2:1. These three bounding boxes are called anchor boxes.

In the mesh based algorithm, it is assumed that the mesh is 19 × 19. If 80 categories of goals are predicted, the training label based on Anchor box is shown in Fig. 7.11. When only one target is detected in each grid, the label dimension corresponding

Fig. 7.10 Anchor box

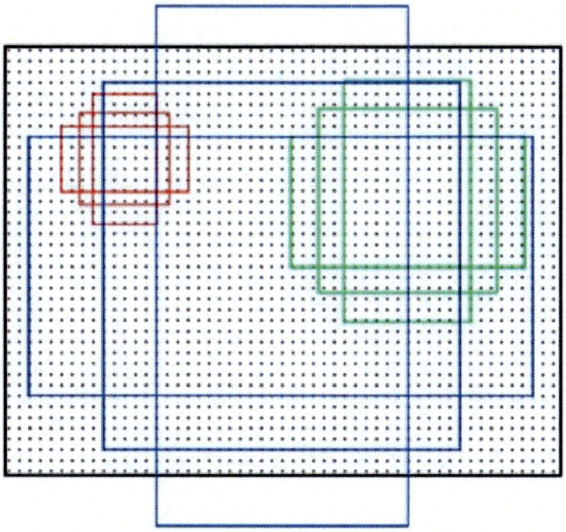

to each feature point is 85 dimensions. In order to detect multiple targets in a grid, set the number of anchor boxes as 3, and the label dimension corresponding to each feature point is $85 \times 3 = 255$, that is, the final dimension of label is 255 and the network output is $19 \times 19 \times 255$.

In fact, both R-CNN series algorithms and YOLO algorithm adopt the idea of Anchor box. By customizing the shape of the target in advance, the training process of the network can be more targeted.

3. Non-maximum suppression

After obtaining the output result of convolution network, it is also necessary to detect the output result to obtain the Bounding box, category and confidence probability of the final target. The detection process of YOLOv2 algorithm is shown in Fig. 7.12. With 7×7 as an example, if the number of Anchor boxes in each grid is 2, the output result is $7 \times 7 \times 2 = 98$ prediction vectors. The detection process is as follows.

(1) Select a category and set all components in the category whose confidence probability is lower than a certain threshold to 0.
(2) Rank all confidence probabilities of this category from large to small.
(3) Use the Non-maximum suppression algorithm to remove the redundant Bounding box of this category.
(4) Repeat steps (1), (2) and (3) for other categories.
(5) Take the maximum value in all categories of each grid and take it as the final detection result.

The Non-maximum suppression algorithm is the core of the whole detection process. In order to discard the redundant Bounding boxes, the concept of IoU is introduced. IoU is used to measure the similarity between the two Bounding boxes. The IoU index is shown in Fig. 7.13. After the introduction of IoU, the Non-maximum

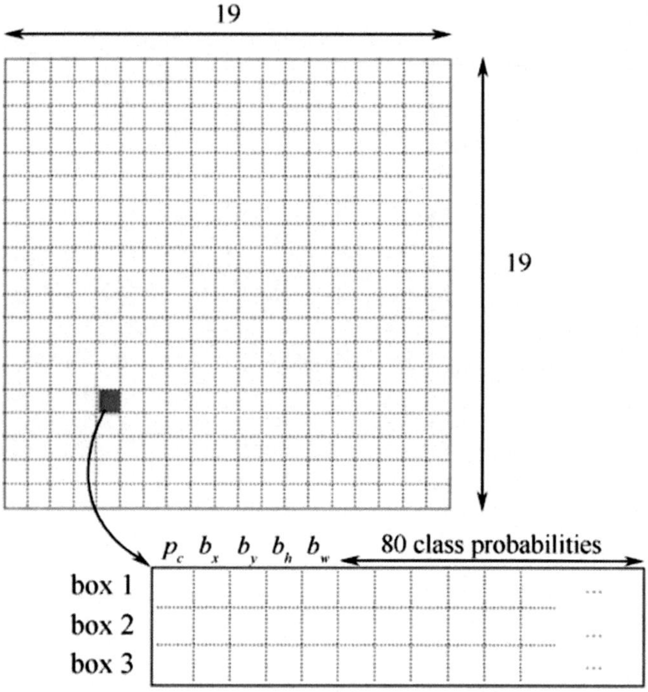

Fig. 7.11 Schematic diagram of training labels based on anchor box

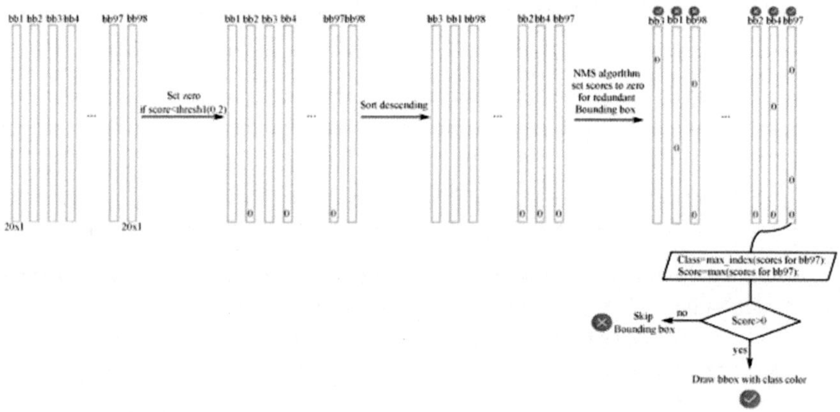

Fig. 7.12 Detection process of YOLOv2

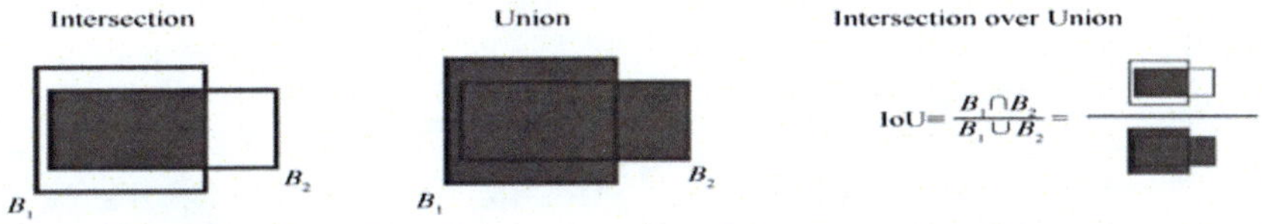

Fig. 7.13 IoU index

suppression algorithm traverses the confidence probability of each grid in the current category from large to small, discards all Bounding boxes with IoU > 0.5, and the remaining Bounding boxes are the Non-maximum suppression results.

7.1.3 Test and Simulation

1. Create a training set

The training set used in this section is 360 images obtained by dividing the video collected by the driving test into frames. The training set is shown in Fig. 7.14.

This book uses the labelImg annotation tool on Github to make training labels, which can be installed using Python's pip command. The tool interface is shown in Fig. 7.15. Click CreateRectBox to mark multiple targets and give the category information, and then click Save to generate the xml file corresponding to the image, as shown in Fig. 7.16, the xml file where xmin, ymin, xmax, and ymax are the pixel information needed to generate training labels.

2. Vehicle inspection results and analysis

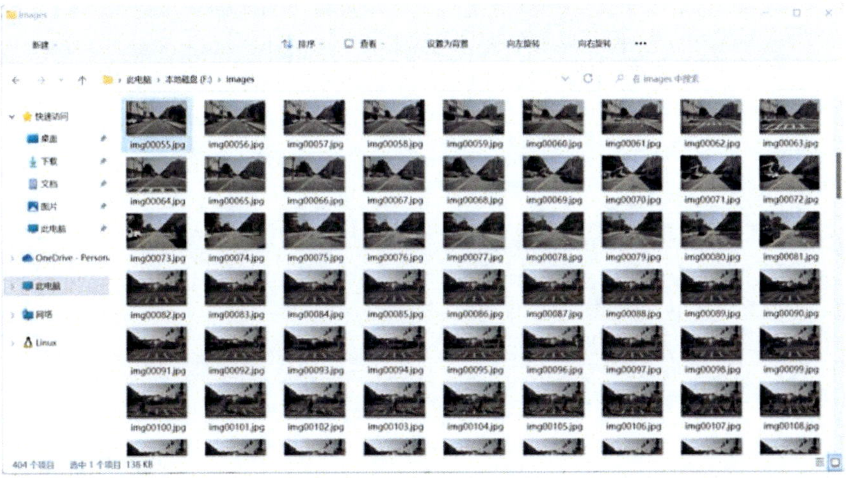

Fig. 7.14 Training set

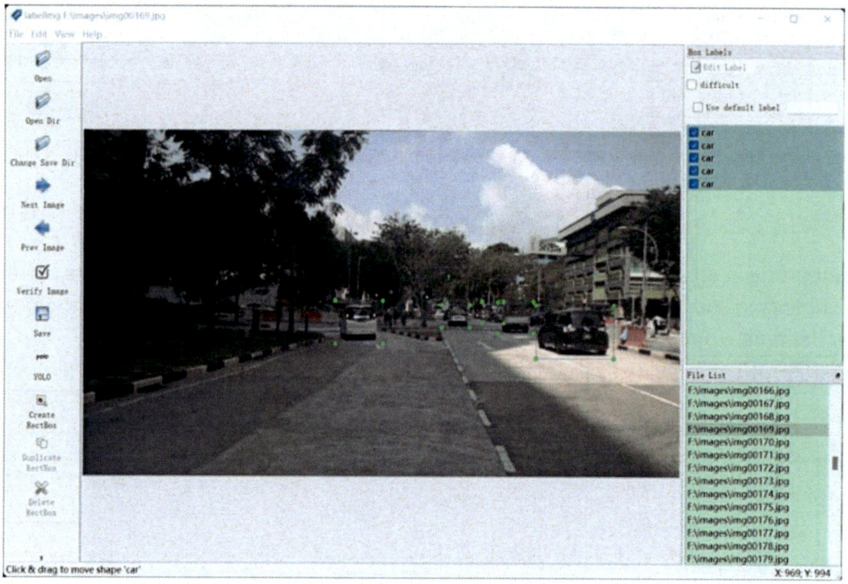

Fig. 7.15 Tool interface

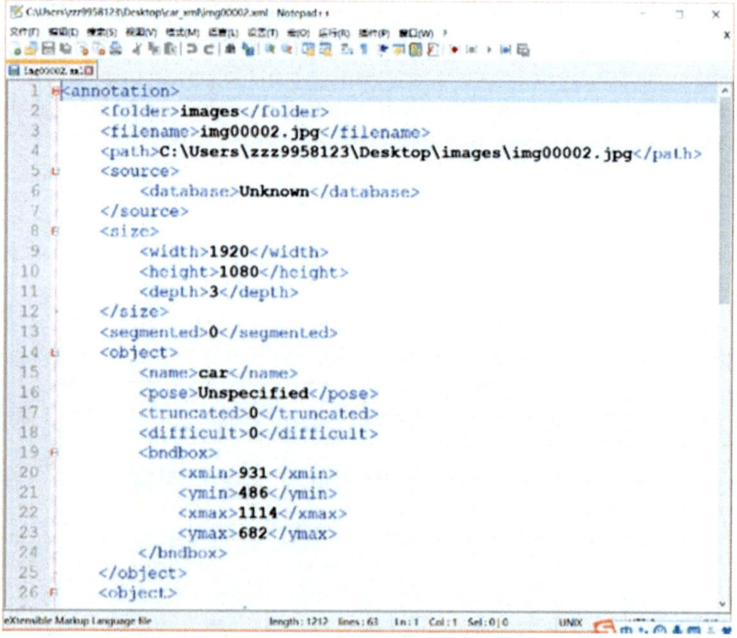

Fig. 7.16 XML file corresponding to the image

This section adopts a training set with 30,000 actual drive test images and a test set with 3000 photos for experiments. The following is an analysis of the detection results of the YOLOv2 algorithm in different scenarios from three aspects.

(1) Vehicle identification

When calibrating the training labels of the training set, the vehicles are divided into three categories: Car, Bus and Truck. The recognition effect of the algorithm for 4 consecutive frames in two scenes is shown in Fig. 7.17. It can be seen from the figure that the algorithm is very robust to targets with a short distance, and only a few frames will identify as Car.

The vehicle recognition results are shown in Table 7.1. It can be seen from Table 7.1 that the recognition rate of close-range targets can reach more than 90%.

(2) Recognition of day and night

The above shows the recognition accuracy of the algorithm during the day. At night, it can be seen from Fig. 7.18 that the recognition effect is the same as that in the daytime. Without considering the model, 957 vehicles were counted, and the algorithm identified 889, with a recognition rate of 92.9%. Therefore, day or night has little effect on the recognition rate of the algorithm.

(3) Small target recognition

Small target recognition is a difficult point in target detection tasks. YOLO algorithm cannot complete small target recognition well. YOLOv2 algorithm used in this chapter integrates high-level features and low-level features by introducing a residual network to achieve long-distance small target recognition. The effect is shown in Fig. 7.19. In the case of only considering small targets at a distance, 578 vehicles were counted, and the algorithm could identify 326, with a recognition rate of 56.4%. Since the YOLO algorithm can hardly be recognized, it is believed that the residual network can effectively solve the problem.

7.2 Information Fusion Based on Millimetre Wave Radar and Machine Vision

7.2.1 Data Fusion of mmWave Radar and Cameras

With the rapid development of smart cars, the research and application of sensor fusion have become a hotspot in academia and business circles. Millimeter-wave radars and cameras are indispensable in automotive collision avoidance warning systems, which are equivalent to eyes. This section adopts a joint calibration method of Millimeter-wave radar and camera. By calibrating the Millimeter-wave radar and camera respectively, the calibration parameters can be obtained. Through these

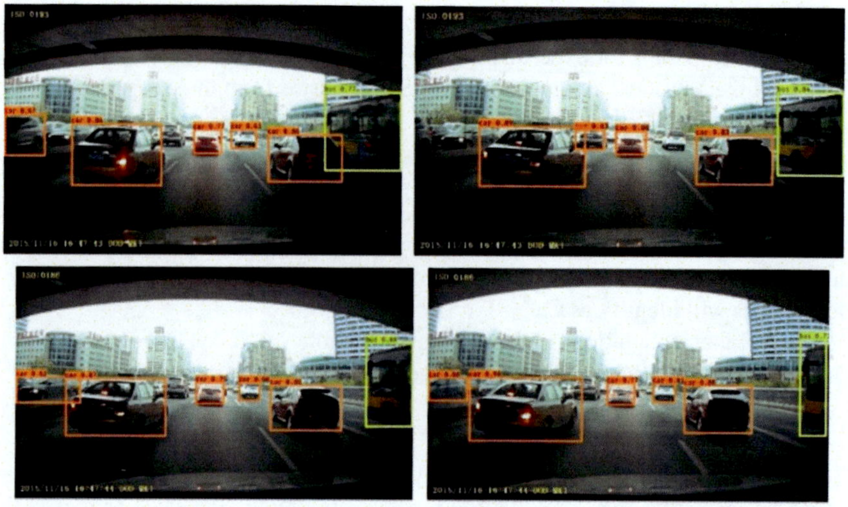

(a) Scenes 1

(b) Scenes 2

Fig. 7.17 The recognition effect of the algorithm for 4 consecutive frames in two scenes

Table 7.1 Vehicle recognition results

Results	Vehicle type		
	Car	Bus	Truck
Total statistics	1983	74	141
Identification number	1872	71	128
Recognition rate	94.4	95.9	90.8

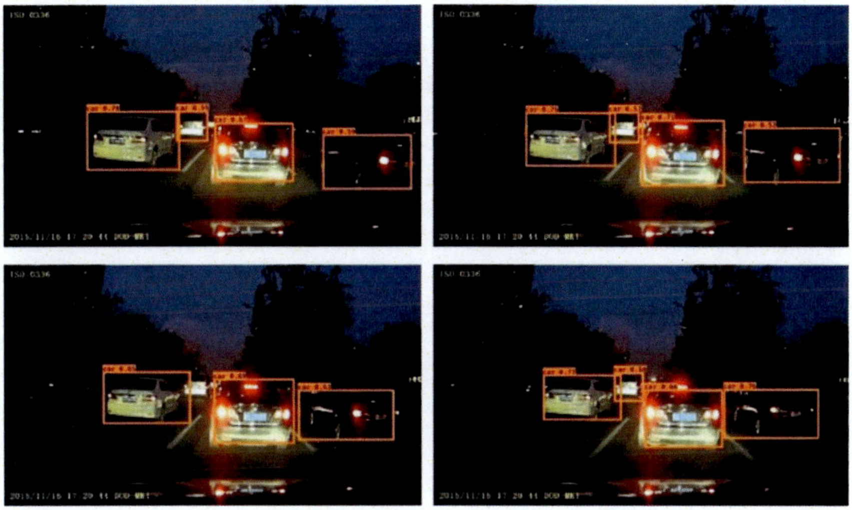

Fig. 7.18 The recognition effect of the algorithm in 4 consecutive frames at night

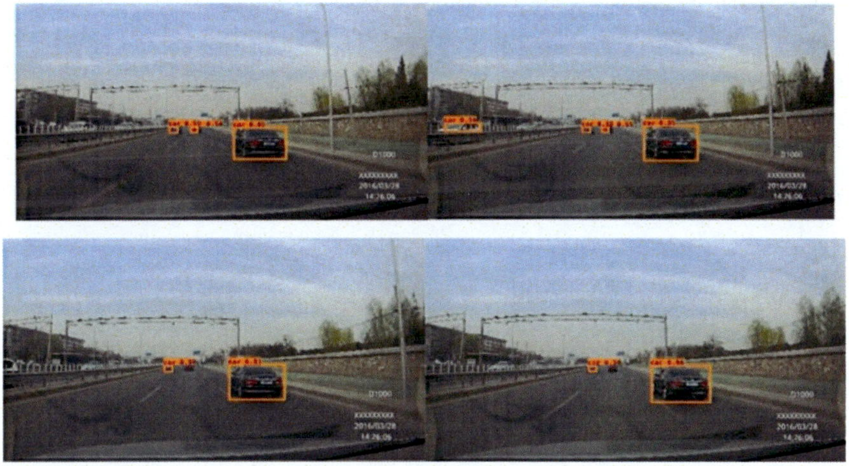

Fig. 7.19 Long-distance small target recognition effect

parameters, the coordinate conversion relationship between the two can be established. The experimental results show that after the coordinate transformation, the target position can be better restored in another kind of data, data fusion can be realized, and reliable data can be provided for the automobile collision avoidance warning system.

1. Camera parameter calibration and distortion calibration

Because a fixed-focus camera is used, the calibration of its built-in parameters is not affected by the environment. This section uses Zhang Zhengyou's calibration method to calibrate the built-in parameters, which are simple to operate and has high precision. Shoot a checkerboard pattern from multiple angles, and use the obtained image as a calibration sample. The calibration sample is shown in Fig. 7.20.

Input the calibration sample into the calibration toolbox of MATLAB, the toolbox can accurately detect all the corner points of the chessboard and automatically complete the calibration of the built-in parameters of the camera. The built-in parameters of the camera are shown in Table 7.2. Among them, the focal length and principal point are used for coordinate transformation, and the distortion parameter is used for image distortion calibration.

The monocular camera used in the experiment will show barrel distortion, and the center of the image will protrude outward. The radial distortion calibration formula is used for calibration.

$$\begin{cases} x' = x + (k_1xr^2 + k_2xr^4 + k_3xr^6) \\ y' = y + (k_1yr^2 + k_2yr^4 + k_3yr^6) \end{cases} \tag{7.6}$$

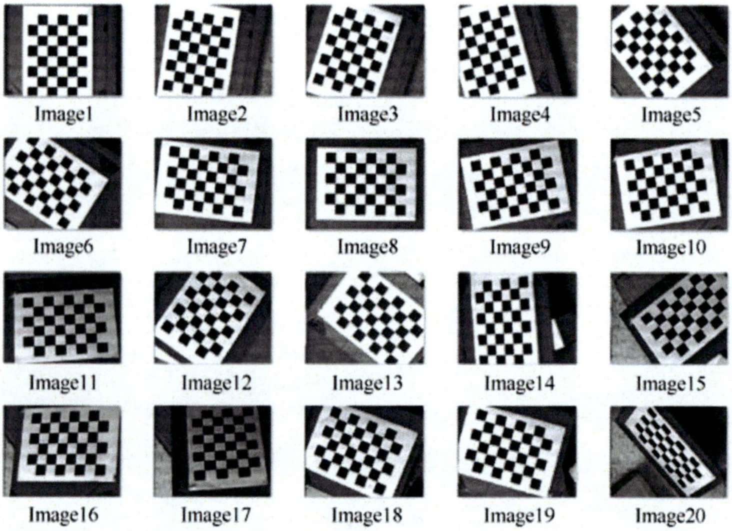

Fig. 7.20 Calibration sample

Table 7.2 Built-in parameters of the camera

Parameter	Calibration value
Focal length	$[f_x, f_y] = [1359.267, 1355.632] \pm [37.584, 35.528]$
Principal point	$[c_x, c_y] = [715.559, 497.772] \pm [7.505, 20.833]$
Distortion	$[k_1, k_2, k_3] = [-0.410, 0.207, 0.000] \pm [0.073, 0.048, 0.000]$

In the formula, x and y are the coordinates before calibration, x' and y' are the coordinates after calibration, $r^2 = x^2 + y^2$.

2. Coordinate conversion of millimeter wave radar and camera

Through calibration, the millimeter-wave radar is parallel to the transverse section of the camera and perpendicular to the ground. On this basis, to ensure that the data of the radar and the camera can be well integrated, it is necessary to establish a projection coordinate system for the radar and the camera respectively. The radar projection coordinate system is $O_{rw} - x_{rw}y_{rw}z_{rw}$, the camera projection coordinate system is $O_{cw} - x_{cw}y_{cw}z_{cw}$, and the x of the two projection coordinate systems is The axes represent the transverse section directions of the radar and the camera, respectively, the y represents the normal direction of the respective transverse sections, and the z represents the vertically upward direction relative to the ground.

The pixel coordinate system $O_p - x_p y_p$ and the camera coordinate system $O_c - x_c y_c z_c$ are established. The x_c axis and the y_c axis in the camera coordinate system are respectively parallel to the x_p axis and the y_p axis in the pixel coordinate system. The relationship of each coordinate system is shown in Fig. 7.21.

The z_c axis in the camera coordinate system is the optical axis direction of the camera. The relationship between the pixel coordinate system and the camera coordinate system is

$$\begin{cases} a = \frac{x_c}{z_c} = \frac{x_p - c_x}{f_x} \\ b = \frac{y_c}{z_c} = \frac{y_p - c_y}{f_y} \end{cases} \tag{7.7}$$

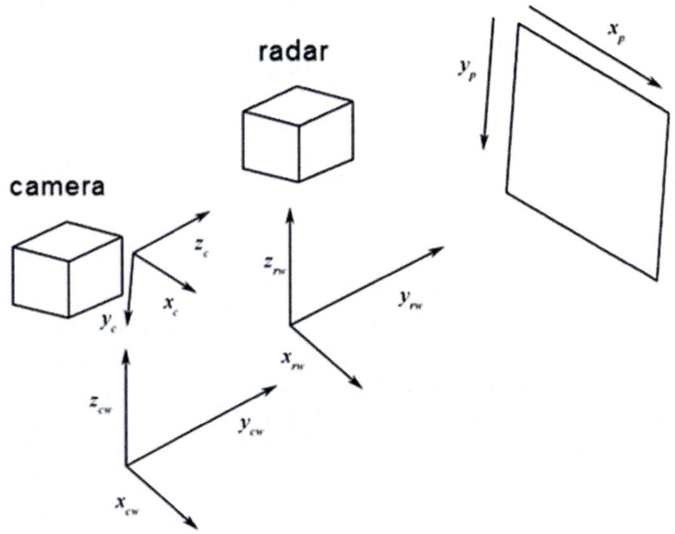

Fig. 7.21 Relationship of each coordinate system

The camera coordinate system $O_c - x_c y_c z_c$ and the camera projection coordinate system $O_{cw} - x_{cw} y_{cw} z_{cw}$ can be converted to each other through translation and rotation, and the conversion relationship is:

$$\begin{bmatrix} x_c \\ y_c \\ z_c \end{bmatrix} = \begin{bmatrix} 1 & 0 & 0 \\ 0 & -\sin\theta & -\cos\theta \\ 0 & \cos\theta & \sin\theta \end{bmatrix} \begin{bmatrix} x_{cw} \\ y_{cw} \\ z_{cw} \end{bmatrix} + \begin{bmatrix} 0 \\ H\cos\theta \\ H\sin\theta \end{bmatrix} \tag{7.8}$$

Assuming that the test site is flat, the coordinates z_{cw}=0 of the ground target in the camera projection coordinate system can be obtained from Eqs. (7.7) and (7.8)

$$\begin{cases} x_{cw} = a y_{cw} \cos\theta + aH\sin\theta \\ y_{cw} = \frac{H(\cos\theta - b\sin\theta)}{b\cos\theta + \sin\theta} \end{cases} \tag{7.9}$$

The radar projection coordinate system $O_{rw} - x_{rw} y_{rw} z_{rw}$ and the camera projection coordinate system $O_{cw} - x_{cw} y_{cw} z_{cw}$ can be converted by simple translation, and the conversion relationship is

$$\begin{cases} x_{cw} = x_{rw} - L_x \\ y_{cw} = y_{rw} + L_x \end{cases} \tag{7.10}$$

By formula (7.8), formula (7.9) and formula (7.10) can be obtained

$$\begin{cases} x_p = c_x + \frac{(x_{rw} - L_x) f_x}{H\sin\theta + (y_{rw} + L_y)\cos\theta} \\ y_p = c_y + \frac{[H\cos\theta - (y_{rw} + L_y)\sin\theta] f_y}{H\sin\theta + (y_{rw} + L_y)\cos\theta} \end{cases} \tag{7.11}$$

In the formula, c_x and c_y are the main points; f_x and f_y are the focal lengths, which are obtained by calibration; L_x is the distance between the x_{rw} axis and the x_{cw} axis in Fig. 7.21; L_y is the distance between the y_{rw} axis and the y_{cw} axis in Fig. 7.21.

3. Data Fusion

With the vehicle movement and the rapid changes in the surrounding environment, the measurement information at different times may be very different. For high-speed moving vehicles, even if the delay time is very short, it will bring a large measurement error. Therefore, the accuracy of data fusion can be ensured only when different sensors detect data at the same time. In this section, the target data is fused in the two dimensions of time and space.

In the time dimension, the sampling period of the millimeter-wave radar used is 100 ms, which means that 10 frames of data can be got per second; the shooting frame rate of the camera is 30 fps, and the sampling period is 33.3 ms. The sampling period of the millimeter-wave radar is 3 times the sampling period of the camera. Therefore, as long as the data acquisition software of the host computer ensures that the millimeter-wave radar and the camera start to collect at the same time, the synchronization of the radar and camera data in time can be realized.

In space, the millimeter-wave radar and the camera have different positional relationships with the target. The position of the target detected by the millimeter-wave radar can be converted to the position of the target in the pixel coordinate system of the camera by formula (7.11).The corner reflector with a strong echo signal is used as the detection target, and its echo signal is relatively concentrated, which is easy to be detected by millimeter-wave radar. The coordinates of the bottom pedestal of the corner reflector are taken as the position coordinates of the target, and the corner reflectors are placed in 9 different positions. The millimeter-wave radar and the camera diagonal reflector are used to collect data respectively. After processing the data, the radar data can be collected. Convert to an image and measure the intuitive effect of data fusion by the position of the radar data in the image. Since the color of the corner reflector is darker, to facilitate viewing, the image is processed by inverse color processing and clipping processing, and the mapping from radar data to an image is obtained as shown in Fig. 7.22.

To measure the accuracy of data fusion, the radar measurement value and the camera measurement value are compared with the real position of the target, and the radar measurement value can be obtained by the target detection algorithm according to the echo signal of the corner reflector; according to the pixel points of the corner

Fig. 7.22 Mapping of radar data to images

Table 7.3 Comparison of measurements for radar and camera

Corner reflector distance		Radar measurements		Camera measurements	
Portrait (m)	Horizontal (m)	Portrait (m)	Horizontal (m)	Portrait (m)	Horizontal (m)
9.13	−1.52	9.02	−1.66	9.21	−1.33
9.13	0.01	9.27	−0.05	9.24	0.06
9.13	1.51	9.33	1.61	8.96	1.65
12.08	−1.49	12.21	−1.57	12.29	−1.43
12.08	−0.02	12.45	0.09	12.23	0.07
12.08	1.52	12.39	1.58	12.31	1.69
14.94	−1.51	15.28	−1.68	15.11	−1.64
14.94	0.02	14.86	0.05	15.17	−0.04
14.94	1.48	15.33	1.46	15.08	1.54

reflector pedestal, The camera measurement value is calculated by the formula (7.9). The comparison between the radar measurement value and the camera measurement value is shown in Table 7.3.

As can be seen from Table 7.3, the difference between the radar measurement value and the camera measurement value and the longitudinal distance of the corner reflector is less than 0.5 m, and the difference between the lateral distance is less than 0.3 m, which meets the data fusion requirements.

7.2.2 *Machine Learning-Based Vehicle Recognition*

As the operating frequency of millimeter-wave radar increases, the radar can find more strong scattering points on the vehicle body. According to the distance and the angle corresponding to these strong scattering points, map them into the actual scene, and the outline of the vehicle body can be obtained, which provides conditions for vehicle identification, and then formulates corresponding warning strategies according to the vehicle type.

Based on the radar signal characteristics, the data set is calibrated by the YOLO algorithm, the target data set is established, and the data set is identified and classified by the machine learning algorithm. Figure 7.23 shows the flow of the model recognition algorithm based on machine learning.

1. Feature extraction

The vehicles large and small have different scattering characteristics and scattering centers, as well as frequency characteristics. The frequency characteristics are shown in Figs. 7.24 and 7.25 respectively.

The echo signal reflected by the scattering point includes not only the echo of the vehicle body but also clutter, which cannot be measured according to the physical

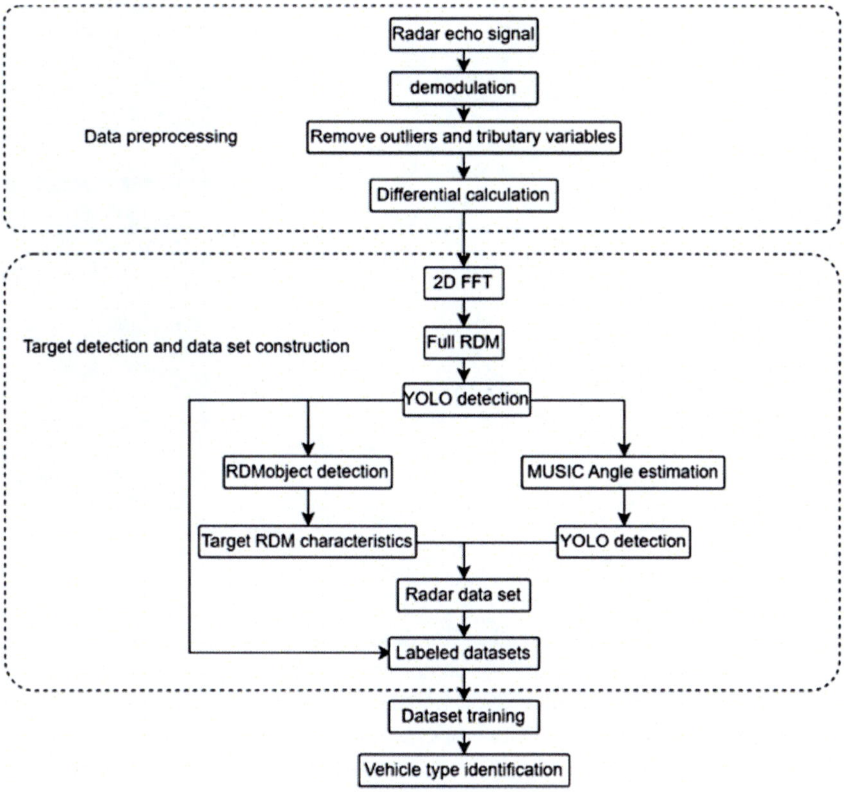

Fig. 7.23 Process of vehicle recognition algorithm based on machine learning

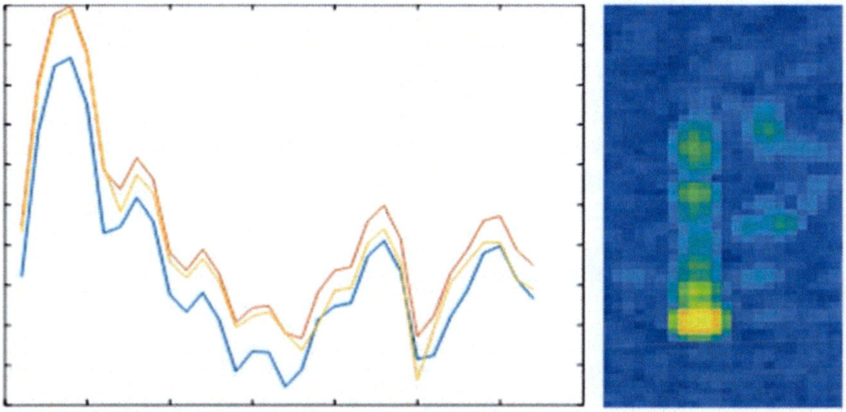

Fig. 7.24 Frequency characteristics of large vehicles

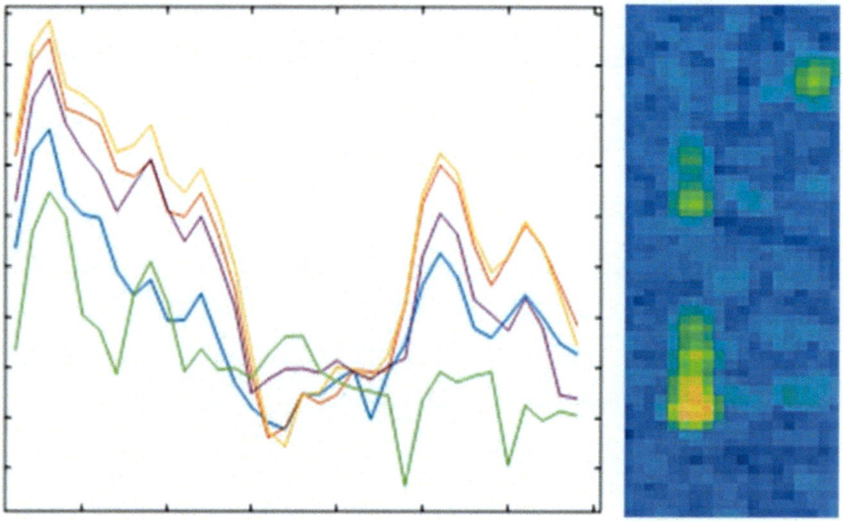

Fig. 7.25 Frequency characteristics of small vehicles

model. Therefore, the differential calculation of echo can reduce the influence of clutter on echo and better reflect the complete information of the vehicle.

To preprocess the received radar echo signal, remove distortion and DC components, and perform the differential calculation. Then perform a two-dimensional Fourier transform to obtain a range-Doppler map (Range Doppler Map, RDM), from which the Doppler spectrum of the target motion can be extracted.

After the above processing, the intuitive characteristics of the signal can be obtained, including distance dimension frequency, velocity dimension frequency, and energy intensity. In addition, it is also necessary to obtain the shape characteristics of the target, so it is necessary to estimate the angle of the target. The music algorithm is used to obtain the estimated spectrum of the target angle, and the multi-frame angle estimation spectrum of the target is accumulated in time according to the moving track of the target in a period to form an angle time map (ATM).

It is difficult to converge because of the large difference in value between RDM and ATM. Therefore, the data needs to be normalized to obtain normalized RDM and ATM. The normalized RDM and ATM are concatenated to obtain the RDM-ATM feature, which is used as the training set.

2. LeNet

Convolutional neural networks (CNN) are mainly composed of convolutional layers, pooling layers, and fully connected layers. The convolutional layer is the core of the CNN, which extracts image features through the convolution kernel (filter); the pooling completes the downsampling operation, reducing the amount of computation and data dimension; the fully connected layer can convert the pooling layer into a vector and perform target classification. The advantage of CNN training is that it can use the error between the output value and the real value for back-propagation, to adjust the weight coefficients to optimize the CNN and further improve the accuracy of forwarding propagation.

Choosing a suitable neural network can greatly promote the accuracy and training speed of target recognition. The classic CNN models include LeNet, AlexNet, GoogleNet, VGG, and DeepResidualLearning. The number of network layers is increasing. The more total layers, the higher the corresponding target recognition accuracy and the better the classification effect. The following problems exist in the application.

(1) The more layers of the network, the more nodes in each layer, and the more parameters that need to be configured. When the training set is limited, the model trained by the network is prone to overfitting.
(2) The more layers, the greater the amount of calculation, the training speed will be greatly reduced, and the overall efficiency will be affected.
(3) The more layers, the more useless nodes appear, and the more difficult it is to optimize the model. Since the task goal is mainly to complete the identification of lines, trends, and peaks, it can be achieved by using a relatively simple convolutional network. Therefore, abandon the use of multi-layer networks such as GoogleNet, VGG, and Deep Residual Learning, and use LeNet. The structure of LeNet is shown in Fig. 7.26.

The first layer of LeNet is the input original image, which is called the input layer. The second and fourth layers and the third and fifth layers are convolutional layers and downsampling layers (also called pooling layers), respectively. The six layers are fully connected layers, which use the softmax function for classification and serve as the output layer of the target.

Convolutional layers are the core of CNN. In image recognition, two-dimensional convolution is usually used. In simple terms, it is to traverse all the positions in the two-dimensional image and calculate the inner product according to the center pixel and its neighboring pixels. For relatively simple, only needing a few different convolution kernels that could extract edge information, line information, and corner features in the image.

Pooling is a form of downsampling. Its basic idea is to retain the main features. Similar to the dimensionality reduction operation of principal component analysis,

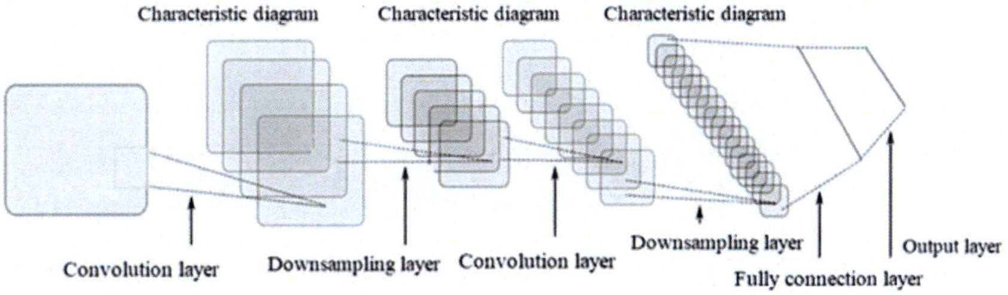

Fig. 7.26 Structure of LeNet

the parameters of the next layer should be reduced to prevent the excessive number of layers from causing too much computation and suppress overfitting.. Usually, a pooling layer is added after the convolutional layer for downsampling. The pooling includes maximum pooling, average pooling, etc. The average pooling is used below.

Regarding RELU as the activation function and uses the Stochastic Gradient Descent (SGD) method. Compared with other gradient descent methods, SGD has a faster training speed and fewer data constraints and can be widely adopted in a variety of data sets.

7.2.3 Experimental Verification and Analysis

In this section, the experiment adopts the forward collision avoidance radar of Beijing Chuansu Microwave Technology Co., Ltd., its operating frequency is 77 GHz, the bandwidth is 4 GHz, and the antenna is an antenna array with 1 transmit and 4 receive. Since there may be relatively few large vehicles in the actual urban road environment, the static verification method is adopted, the radar and the camera are fixed by two tripods respectively, and the calibration is carried out according to the joint calibration method. Go to the vehicle to collect data and build a dataset.

1. Create a training set

Radar data can be obtained by a target detection algorithm, but it does not contain vehicle type information, so it is necessary to supplement the data with image data. Taking the traffic light intersection as the experimental scene, the data from radar and camera are collected synchronously for the incoming and outgoing targets respectively. The target collection scene is shown in Fig. 7.27.

Take raw data and process it into target data. Use the YOLO algorithm to detect and identify images, and the effect is shown in Fig. 7.28.

The YOLO algorithm can quickly and effectively detect and identify targets, and output their pixel coordinates. According to the pixel coordinates, the actual position of the target can be calculated using Eq. (7.11). Then, the actual position of the target is mapped to the radar coordinate system, so that the horizontal and vertical distances of the target are converted into the radial distance from the target to the radar. Assuming that the speed of the target does not change drastically when the speed is less than 100 ms, the radial speed of the target is deduced from the position

Fig. 7.27 Target collection scene

Fig. 7.28 Detection and recognition effect

change of the multi-frame target, and the target is located in the distance-velocity two-dimensional map of the radar echo signal. Map the complete frontmost vehicle shown in Fig. 7.27 to the radar speed-distance two-dimensional map, and the target positioning result in the speed-distance two-dimensional map is shown in Fig. 7.29.

Extracting The 30×9 area of the target's position in the image to obtain the RDM, then the ATM is obtained according to several frames of continuous data to form the RDM-ATM feature, and the feature extraction of the radar data is completed. The target classification result in the image is used as the classification label of the training data to obtain a complete dataset.

2. Target recognition results

The training set used in the experiment consists of single-target data of 600 small cars and 400 large cars, and the test set consists of single-target data of 100 small cars and 100 large cars.

Setting the maximum number of training epochs to 3, a training epoch trains the entire training set for a full cycle, and each epoch of training shuffles the data. The training initial learning rate is set to 0.01, and the number of iterations is 21, with 7 times per round. The rate of learning and loss are shown in Fig. 7.30 and 7.31, respectively.

The dataset can be classified using a SVM classifier and random forest classifier based on principal component analysis. The classification effects of several classification methods are shown in Table 7.4.

As can be seen from Table 7.4, the classification results of the SVM classifier, random forest classifier and LeNet are all very good. The effectiveness of the target recognition method is verified.

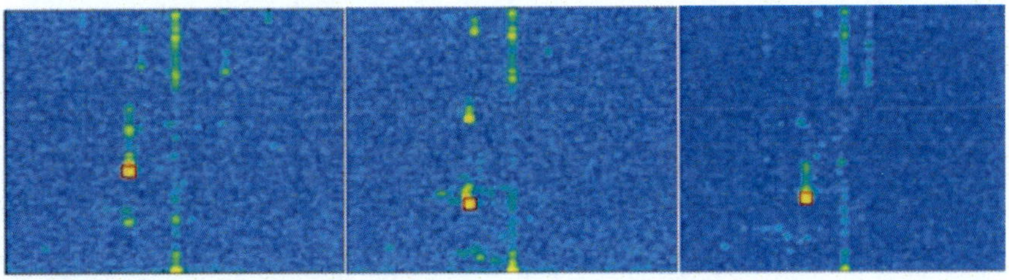

Fig. 7.29 Target positioning results in the speed-distance 2D map

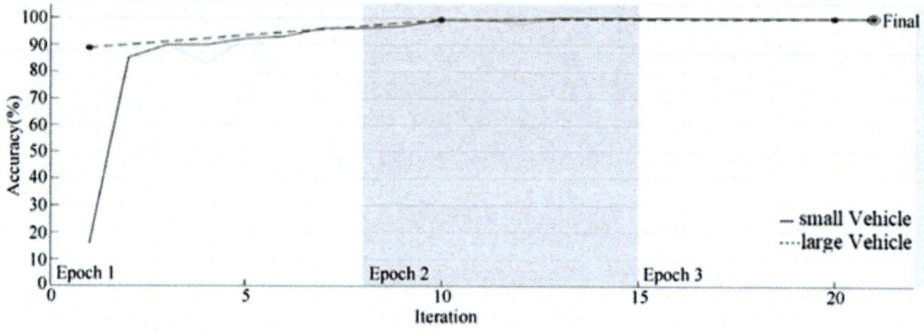

Fig. 7.30 Learning rate

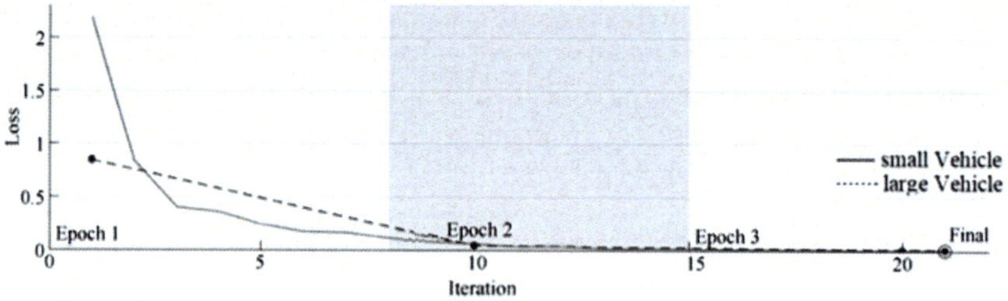

Fig. 7.31 Rate

Table 7.4 Comparison of classification effects of several classification methods

Classification	Recognition rate of small car (%)	Recognition rate of large vehicle (%)
SVM	93	91
Random forest	91	87
LeNet	97	99

7.3 A Target Detection Method Based on the Fusion Algorithm of Radar and Camera

From the perspective of sensor fusion, this paper aims to reduce the complexity of the fusion algorithm and improve the real-time performance of target detection. The fusion of radar information and camera information is realized by fusing EPNP algorithm and center fusion algorithm, the radar detection is connected with the target center point. Using radar features and image features, the initial detection can be improved by re-estimating the target's depth, line of sight, rotation, and attributes. In the fusion stage, the original EPNP algorithm needs a lot of time-consuming calculation of matrix coordinates when calculating space coordinates. Therefore, an improved image association method is added to speed up the algorithm. When using the center fusion algorithm, the original frustum method is improved to the truncated cone method. Mapping, image method detection in the pre-processed radar signal area, greatly reducing the time required for target detection. At the same time, due to the addition of Gauss–Newton optimization method in EPNP algorithm, this fusion method solves the common overlap problem in the fusion method. The entire algorithm is described in detail below.

In this part, we will introduce our improved method, the sensor fusion method of radar and camera. The algorithm flow chart is shown as in Fig. 7.32. We link radar detection with the target center point, using radar features and image features, and re-estimating the depth of the target. Sight distance, rotation and attributes to improve initial detection. First using the EPNP algorithm to determine the center point of the object on the image plane, returning to other object attributes, such as position, direction, distance. We combine these center points with the translation matrix, rotation matrix, camera pose, etc. Through the center fusion method improve the radar information on the image, and finally achieve more comprehensive target detection.

For each detection, the radar also reports the instantaneous velocity of the object in the radial direction. This radial velocity does not necessarily correspond to the actual velocity vector in the direction of movement of the object. For target A, the speed and radial speed in the vehicle coordinate system are the same v^A. On the other hand, for target B, the position of the object in the vehicle coordinate system reported by the radial velocity radar v^r is different from the actual velocity v^B. Figure 7.33 illustrates the difference between the reported radial direction of the radar and the actual velocity coordinate system of the object in the vehicle.

We use superscripts w and c respectively to indicate the coordinates in the world coordinate system and the camera coordinate system. Then the coordinates of the 3D reference point in the world coordinate system are $P_i^c, i = 1, \ldots, n$. The coordinates of the 4 control points in the world coordinate system are $c_j^w, j = 1, \ldots, 4$, The coordinates in the camera reference coordinate system are $c_j^c, j = 1, \ldots, 4$. The reference point projection map is shown in Fig. 7.34.

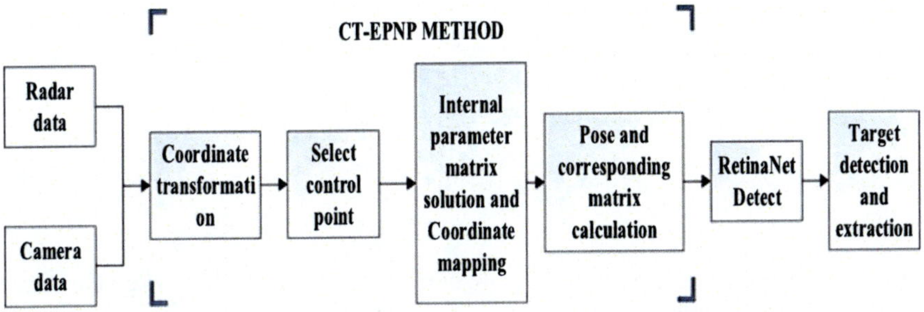

Fig. 7.32 CT-EPNP Algorithm flowchart

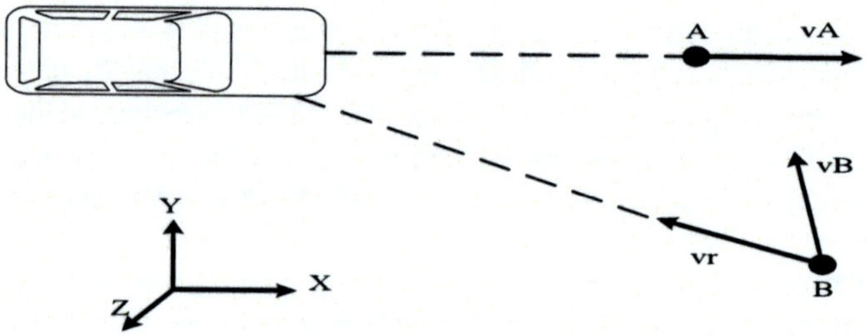

Fig. 7.33 Difference between actual speed and radial speed

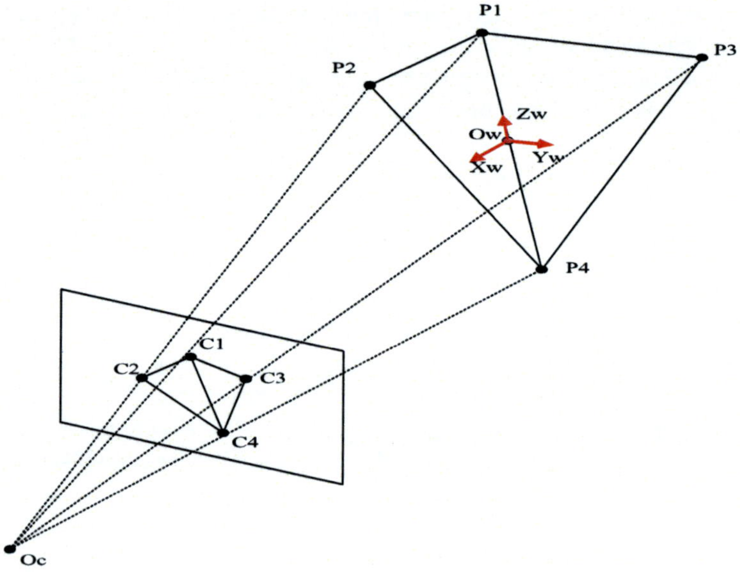

Fig. 7.34 The projection of the reference points

The EPNP algorithm expresses the coordinates of the reference point as a weighted sum of the control point coordinates:

$$P_i^w = \sum_{j=1}^{4} \alpha_{ij} c_j^w, with \sum_{j=1}^{4} \alpha_{ij} = 1 \tag{7.12}$$

where α_{ij} is the coordinate of the homogeneous center of gravity. When the virtual control points are determined, the premise that the four control basic points are not coplanar is satisfied. α_{ij}, $j = 1, \ldots, 4$ c is the only certain. In the camera coordinate system, the same weighted sum relationship exists:

$$P_i^c = \sum_{j=1}^{4} \alpha_{ij} c_j^c \tag{7.13}$$

Jesse´s Comment: Assume that the external parameters of the camera are $\begin{bmatrix} R \ \mathrm{t} \end{bmatrix}$. Then there is a relationship between virtual control points c and f:

$$c_j^c = \begin{bmatrix} R \ \mathrm{t} \end{bmatrix} \begin{bmatrix} c_j^w \\ 1 \end{bmatrix} \tag{7.14}$$

The EPNP algorithm expresses the coordinate of the reference point as the weighted sum of the coordinate of the control point, which can be obtained:

$$P_{\mathrm{i}}^c = \begin{bmatrix} R \ \mathrm{t} \end{bmatrix} \begin{bmatrix} P_i^w \\ 1 \end{bmatrix} = \begin{bmatrix} R \ t \end{bmatrix} \begin{bmatrix} \sum_{j=1}^{4} \alpha_{ij} c_j^w \\ 1 \end{bmatrix} \tag{7.15}$$

$$P_{\mathrm{i}}^c = \begin{bmatrix} R \ \mathrm{t} \end{bmatrix} \begin{bmatrix} \sum_{j=1}^{4} \alpha_{ij} c_j^w \\ \sum_{j=1}^{4} \alpha_{ij} \end{bmatrix} = \sum_{j=1}^{4} \alpha_{ij} c_j^c \tag{7.16}$$

At the same time, suppose we reduce one control point. Under the condition of three control points, assuming that the requirement $\sum_{j=1}^{3} \alpha_{ij} = 1$ is satisfied, we can get:

$$P_{\mathrm{i}}^w = \begin{bmatrix} x_i^w \\ y_i^w \\ z_i^w \end{bmatrix} = \begin{bmatrix} c_1^w \ c_2^w \ c_3^w \end{bmatrix} \tag{7.17}$$

And need to satisfy $\sum_{j=1}^{3} \alpha_{ij} = 1$, there are 4 equations in total, 3 unknowns, only solutions in the sense of least squares exist. In summary, 4 basic points of non-coplanar control are required.

Here we take the reference point set $\left\{ P_{\mathrm{i}}^w, i = 1, \ldots, n \right\}$, and select the center of gravity of the reference point as the first control point:

$$c_1^w = \frac{1}{n} \sum_{i=1}^{n} P_{\mathrm{i}}^w \tag{7.18}$$

$$\text{Get matrix} : A = \begin{bmatrix} P_1^{w^T} - c_1^{w^T} \\ \cdots \\ P_n^{w^T} - c_1^{w^T} \end{bmatrix} \tag{7.19}$$

Let the characteristic value of $A^T A$ be $\lambda_{c,i}$, $i = 1, 2, 3$ The corresponding feature vector is $v_{c,i}$, $i = 1, 2, 3$. Then the remaining three control points can be determined according to the following formula:

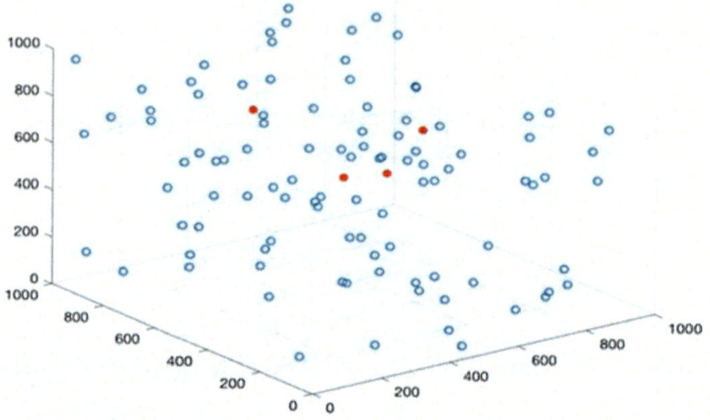

Fig. 7.35 The selection of control points for CT-EPNP

$$c_j^w = c_1^w + \lambda_{c,j-1}^{\frac{1}{2}} v_{c,j-1},\ j = 2, 3, 4 \tag{7.20}$$

The following Fig. 7.35 is examples of the selection of CT-EPNP algorithm control points. Blue is reference points, red is 4 control points, and the reference coordinate system is the world coordinate system.

Coordinate mapping

Let K be the internal parameter matrix of the camera, which can be obtained through calibration. $\{u_i\}_{i=1}, \ldots, n$ is the reference point. $\{P_i\}_{i=1}, \ldots, n$ is the mapping on the image:

$$\forall i, w_i \begin{bmatrix} u_i \\ 1 \end{bmatrix} = K P_i^c = k \sum_{j=1}^{4} \alpha_{ij} c_j^c \tag{7.21}$$

Substitute $c_j^c = \begin{bmatrix} x_j^c\ y_j^c\ z_j^c \end{bmatrix}^T$ into the above formula, write K as focal length f_u, f_v and optical center (u_c, v_c) as follows

$$\forall i, w_i \begin{bmatrix} u_i \\ v_i \\ 1 \end{bmatrix} = \begin{bmatrix} f_u & 0 & u_c \\ 0 & f_v & v_c \\ 0 & 0 & 1 \end{bmatrix} \sum_{j=1}^{4} \alpha_{ij} \begin{bmatrix} x_j^c \\ y_j^c \\ z_j^c \end{bmatrix} \tag{7.22}$$

Corresponding equations:

$$k\sum_{j=1}^{4}\alpha_{ij}f_u x_j^c + \alpha_{ij}(u_c - u_i)z_j^c = 0$$
$$k\sum_{j=1}^{4}\alpha_{ij}f_v y_j^c + \alpha_{ij}(u_c - u_i)z_j^c = 0 \quad (7.23)$$

Connect all n points in series to get a system of linear equations:

$$M\vec{x} = 0 \quad (7.24)$$

Among them $\vec{x} = \left[c_1^{c^T}\ c_2^{c^T}\ c_3^{c^T}\ c_4^{c^T}\right]$, $\vec{x}$ is the vector whose coordinates of the control point in the camera coordinate system are 12×1.

Calculate the coordinates of the control point in the camera reference coordinate system:

$$c_i^c = \sum_{j=1}^{N}\beta_k v_k^{[i]}, \quad i = 1, \ldots, 4 \quad (7.25)$$

Calculate the coordinates of the reference point in the camera reference coordinate system:

$$P_i^c = \sum_{j=1}^{4}\alpha_{ij}c_j^c, i = 1, \ldots, n \quad (7.26)$$

Calculate the center of gravity $\left\{P_{\mathrm{i}}^{w}\right\}_{i=1,\ldots,n}$ of P_0^c and the matrix B:

$$P_0^c = \frac{1}{n}\sum_{i=1}^{n}P_i^c$$
$$B = \begin{bmatrix} P_1^{c^T} - P_0^{c^T} \\ \cdots \\ P_n^{c^T} - P_0^{c^T} \end{bmatrix} \quad (7.27)$$

Calculate the translation matrix H:

$$H = B^T A \quad (7.28)$$

If $|R| < 0$, then $R(2, :) = -R(2, :)$ Calculate the translation t in the pose:

$$t = P_0^c - RP_0^w \quad (7.29)$$

After correlating the radar detection with the corresponding target in the image through the EPNP algorithm, we use the depth and speed of the radar detection to create complementary features for the image. Usually, for each radar detection related to an object, we generate a heat map channel in the corresponding image, located in the center and inside of the 2D image, controlled by the parameter α. The heat map value is the normalized object depth (d), and also the x and y components of the radial velocity (v_x and v_y) in the egocentric coordinate system:

$$F^j_{x,y,i} \frac{1}{M_i} \begin{cases} f_i & \left|x - c^j_x\right| \le \alpha w^j \\ & \left|y - c^i_y\right| \le \alpha h^j \\ 0 & otherwise \end{cases} \tag{7.30}$$

where $i \in 1, 2, 3$ is the feature mapping point, M_i is the normalization factor, f_i is the image feature value, c^x_j and c^y_j are the x and y coordinates of the center point of jth on the image, w^j and h^j are the width and height of jth. If two objects have overlapping heat map areas, then the object with the smaller depth value dominates, because only the closest object is fully visible in the image. This also avoids inaccurate recognition of overlapping objects to a certain extent.

We conducted simulations through Python3.8, and the data set used VOC and nuScenes, which were applied to target detection under the RetinaNet model. A certain degree of improvement has been achieved in accuracy. At the same time, in order to verify the relationship of the radar image mapping, we also performed simulations on the nuScenes data set. The relationship between 3 and 2D mapping is verified. Figure 7.36 shows the result of target detection.

We then extract the detected targets in the figure as shown in Fig. 7.37.

It is not difficult to see that there are the following problems in the extraction of the target. First, the overlapping objects are well resolved. The motorcycle in the above picture and the car in the back picture overlap but both achieve detection and extraction, but there are also some problems. For example, the tricycle detection and

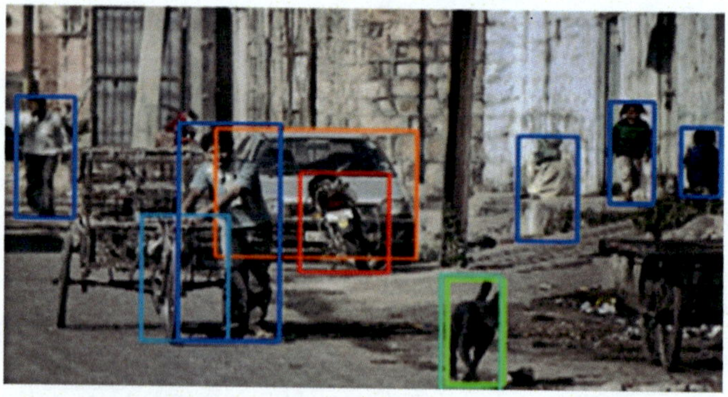

Fig. 7.36 Target detection map

Fig. 7.37 Detected target extraction map

extraction of pedestrians on the left side of the image is divided into two parts for detection and extraction, but in reality it should be considered as an object. Secondly, the black dog at the bottom of the image was detected twice because of the report printed in python. It was tested twice as an animal and other obstacles, But after the radar-camera mapping, it is extracted as a target by the mapping algorithm, which also shows that our improved algorithm has a certain effect in preventing repeated detection. In order to verify the above problems, we have also conducted experiments on the challenging nuScenes data set to verify the effect of our vehicles in the driving process. While considering the accuracy of the algorithm, it is necessary to consider the efficiency of the algorithm. Figure 7.38 is a diagram of the detection effect during the driving of the vehicle.

Fig. 7.38 Detection effect diagram during vehicle driving

7.4 Summary

This chapter uses deep learning algorithms to accurately identify multiple vehicle objects from videos. (1) The two mainstream structures in the current convolutional neural network are introduced and used as the basic module of the network model. (2) The core idea of using the YOLO algorithm for target detection is analyzed. (3) Drawing on the idea of the YOLOv2 algorithm, build a vehicle detection model, and give the algorithm flow of detecting targets from the output results of the model. (4) The establishment of the training set and the generation of training labels are introduced, and the recognition effect of the model is analyzed from three aspects.